THE SUPREME COURT
IN THE FEDERAL
JUDICIAL SYSTEM

THE SUPREME COURT
IN THE FEDERAL
JUDICIAL SYSTEM

Stephen L. Wasby

Southern Illinois University at Carbondale

Holt, Rinehart and Winston

New York   Chicago   San Francisco   Dallas
Montreal   Toronto   London   Sydney

To my grandmother,
    Frances Bunshaft,
    and to the memory of my grandfather,
    Benjamin Bunshaft:

The good people do is not interred with their bones,
but lives in the hearts of those who cherish their memory.

Library of Congress Cataloging in Publication Data

Wasby, Stephen L
  The Supreme Court in the Federal judicial system.

  Bibliography: p. 243.
  Includes index.
  1. United States. Supreme Court. 2. Courts—United
States. I. Title.
KF8742.W38   347'.73'26   74-30706
ISBN 0-03-030426-1
ISBN 0-03-038226-2 pbk.

Printed in the United States of America

890   090   987654321

PREFACE

The United States Supreme Court receives much attention and adulation as well as much criticism. Yet the American public seems to know far less about it—whether about the number of justices or about how the Court functions—than it does about the Congress or the President. Some of this results from the fact that the Court does not engage in "public relations" activities, while the president and our national legislators, particularly the president, have ways of making themselves highly visible. The Court is content to issue decisions of great import for our national life and then to "go about its business." What the average American understands about the Supreme Court depends to a particularly large degree on what others explain about the institution. Yet there has been little public education about how the Court functions or its place in our government, because of the "mystery of the law," because those in the media and our educational system, like most Americans, themselves know relatively little about the Court, and because lawyers who do know, do not make it their task to inform their fellow citizens. Although the "court watcher" who looks closely can find out much that escapes the quick, passing glance, all this reinforces the Court's own secretiveness about its internal processes. The scholars who have devoted their attention to the Court thus provide much of the basis for what appears here.

To understand the Supreme Court, one must know about its internal procedures—how it accepts cases for review and decides those cases—and about its role at the top of our nation's judiciary, that is, how it fits into our system of separate national and state courts—a "dual court system." This requires that we look at the entire federal court system and its personnel, as well as at the relations between federal and state courts and at the administration of the courts—not at the Supreme Court alone. Yet one must also know about the Court's general place in the American system, about its role vis-à-vis our national legislative and executive branches and vis-à-vis state governments, and about the effect of the Court's actions. All of these matters are discussed in this book.

The emphasis throughout is on the Court's operation, not on the doctrine it announces. However, the Court's own cases are used as illustrations and particular attention will be given to the its "doctrine of procedure"—the rules on access to the courts—and to its rulings on actions of Congress and the president. My intent is to provide background for understanding further actions of the Court as well as what it has done so far, because its doctrine may change but its basic way of operating does so much less. The book is divided into nine chapters, grouped into four parts. The first part deals with the Supreme Court's overall role in the American governmental, legal, and political system —including discussion of its power of judicial review and the Court's place in public opinion. The second part covers the structure and administration of the federal court system and the selection of the judges of the federal courts. The third part includes examination of the rules for getting into court and some of the considerations that underlie appeals, followed by treatment of the Court's exercise of choice in determining what cases to decide and the ways the judges behave in deciding cases. The fourth and last part provides an examination of the Court's review of congressional and executive branch actions and of the communication of decisions to those to be affected and the impact of some of those rulings.

Although some data on the Court has been prepared for this book, it is basically a work of synthesis, drawing on studies carried out by others. Thus, emphases in the existing literature will to some extent be reflected in what is presented, as will some of the frequent gaps that exist in the scholarly coverage of our judicial system. The most recent studies have been used as much as possible, and an effort has been made to present salient information through the end of the Supreme Court's 1975 Term (ending in July 1976), with a few important developments after that date also reported. The reader should keep in mind, however, that detailed analyses of the last few years usually do not appear immediately but take time to prepare. I have also drawn on my own prior work, particularly concerning the communication and impact of court decisions, but, it is hoped, without inflicting too much of my own material on the reader; the reader who wishes can pursue those materials more fully at his or her leisure.

In a work of this sort, acknowledgments are many. Over time many people have provided ideas. They include other scholars of the Court,

generous in sharing their thoughts; professional colleagues with whom
I have worked, particularly Joel Grossman of the University of Wisconsin at Madison and Anthony D'Amato of the School of Law, Northwestern University; students for whom I have tried to refine and clarify
ideas; and my department and its chairpersons—Randall Nelson and
John Baker—who helped provide an atmosphere both stimulating of
ideas and sufficiently quiet that one could develop them without constant interruption. Particular thanks go to Denise Rathbun at Praeger
Publishers, to whom I owe particular debts for suggesting that I write
the book and for her comments on the manuscript, and to a special
person in my life, who has probably heard more about this project than
either of us care to say. The Oregon coast again provided the opportunity to think and write and to clear my head by walking on its long
stretches of ocean beach. I take the responsibility for what I have done
with the ideas and the opportunity provided by them; I hope the reader
will receive some profit from what is presented here.

S. L. W.

CONTENTS

45125

PART ONE | THE SUPREME COURT'S ROLE

The Supreme Court of the United States is seen in many different ways by the American public. To some the Court is lawgiver, the nation's High Tribunal and protector of the Constitution—and perhaps upholder of our liberties. To others, the Court is not the finder of the law but the maker of law, a policy maker perhaps usurping the legislative prerogative. Even more negatively, the Court may be seen as an obstacle to change whose members are "nine old men." Whatever the competing views which people hold and regardless of what value, positive or negative, they give to the Court and its performance, many people see the Supreme Court as different from other courts—both other courts in the United States and the highest courts of other nations. Until after World War II, ours was the only judicial system in which the courts had the power of judicial review—the power to invalidate acts of other branches of government as violating the Constitution. Courts in a number of other nations now have that power. However, in comparison with the highest courts of many nations, our Supreme Court remains unusual in not being solely a "constitutional court"

1

but in handling all types of litigation. Thus not only do cases involving statutory interpretation or questions of the Constitution's meaning come before it but so do cases involving the application of already developed policy to particular fact situations ("norm enforcement"), such as its determinations in employee injury cases under federal law. Another major way in which the Supreme Court differs from most other courts, which must take most cases brought before them, is in being able to control the cases it will hear. Perhaps most important in making the Supreme Court an exceptional court is the simple fact that it sits atop the judicial system not only of the national government but of the judicial systems of all the states. Thus, although political action can be taken to reverse or limit what the Court has done, there is no other court to which one can appeal. The Court's ability to choose its cases coupled with its paramount position gives its rulings the widest possible influence.

People expect the Supreme Court to act like a court—or like what they think a court is. Indeed, like any other court, the Supreme Court acts within a particular legal system—in the United States, a legal system supposedly adversary in nature in which judges are expected to rely on precedent. Yet people react to it not only in terms of its procedures but also in terms of its results, treating it like a policy maker. In one sense, all courts are policy makers, even when they seem to be doing little more than recording the actions of others, as when a trial judge accepts a guilty plea worked out between the prosecutor and defense counsel. Yet the Supreme Court's policy making, primarily through its interpretation of statutes and the Constitution, is far more visible than that engaged in by many lower courts even though there are many important policy questions in the cases decided there, for example in the commercial cases which constitute a larger percentage of the dockets of state supreme courts. The potential applicability of the Supreme Court's cases to the entire nation also makes most of its rulings of far greater policy relevance than those of most other courts. The public reaction to the Court's policy making forces the Court, no matter what the mythology about what it does, to become an actor in the political system. Thus the justices must engage in political acts in the broad sense: they must take the Court's environment into account if the Court's actions are to have an effect and if, in the long run, the Court is to survive.

That people look at the United States Supreme Court as both a law court and a political actor requires that particular attention be given to the Supreme Court's role—the topic of the first part of this book. In Chapter 1, in order to provide some historical context, we first look at the Court's relative centrality in American life and the related question of our dependence on the Supreme Court and courts generally and at transitional periods in the Supreme Court's history, particularly the transition from the Warren Court to the Burger Court. Then we turn to a discussion of the Court's role as policy maker, in which we look at the Court's use of "neutral principles" and precedent to decide cases, the Court's "activism" or "self-restraint," and the Court's use of strategy to achieve its goals. The first part of Chapter 2 deals with the development of judicial review and the relation of judicial review to democracy as well as with the question whether the Court has been a check on other policy makers or part of the "dominant national alliance." In the second part of the chapter, the Court's place in public opinion is explored. Several themes repeat themselves through these two chapters. One is the myth that the Court only finds the law, a myth countered by the recognition that the Court plays its most important role as a political institution. Other themes are the Court's place in legitimizing, as well as invalidating, policy, and the support—or lack of it—provided for the Court by elites and other members of the public.

1 | THE SUPREME COURT AS POLICY MAKER

THE PRESENT COURT AND THE HISTORICAL PATTERN: THE COURT'S RELATIVE CENTRALITY

The end of the nation's second century saw a time of change with respect to the Supreme Court. There were changes in personnel. Chief Justice Earl Warren and Justice Abe Fortas left the Court in 1969, followed in 1971 by Hugo Black and John Marshall Harlan and in 1975 by William O. Douglas, on the Court longer than anyone else in its history. Warren Burger became Chief Justice and was joined first by Harry Blackmun, then by Lewis Powell and William Rehnquist, and later by John Paul Stevens. These changes in personnel brought changes in doctrine and in the Court's relationship to public opinion. The new Chief Justice's greater attention to administrative matters, coupled with increasing caseload, also helped to produce the first serious structural examination of the federal court system since the 1920s. The doctrinal changes and the Court's stricter position as to the types of cases which would be heard by the courts made it much less a civil liberties "beacon light" than the Warren Court. Civil liberties activity, including the drawn-out effort to obtain ratification of the Equal Rights Amendment (ERA), was mounted on other fronts. Events surrounding Watergate further deflected attention from the Court toward the president and Congress. These events and the Court's more restrained policy position made it a less central political actor than in the 1950s and most of the 1960s—although still more central than the Court from 1937 to the time when Earl Warren became Chief Justice. Political activity to eliminate or reduce "forced bus-

4

ing" and to limit abortions—activity provoked by the Court's own policies —indicated that the Court had by no means left the arena.

The recent controversy surrounding the Court makes it easy to forget that the Supreme Court's position in our political constellation has changed often. At times, the Court has seemed central to the nation's development; at other times, it has been more in the background. One of the periods of the Court's greatest centrality came in the early nineteenth century under Chief Justice John Marshall, who helped the Court achieve its earliest prominence. Through Marshall, the Court established its power to invalidate acts of Congress. Through decisions such as *Gibbons v. Ogden*, establishing Congress's power over interstate commerce broadly defined, and *McCulloch v. Maryland*, striking down a state tax on the National Bank, the Court played a major role in establishing the national government's place in our scheme of federalism.* Under Chief Justice Roger Taney, the Supreme Court was perhaps less obtrusive. In that period, the Court allowed the states more authority over commerce, for example, ruling in *Cooley v. Board of Port Wardens* that, where diversity was needed, states could act where the federal government had not done so. Unfortunately for Taney's reputation, in the *Dred Scott* case—the Court's second exercise of judicial review at the national level—the Court seriously exacerbated the tensions which led to the Civil War by upsetting Congress's regulation of slavery in the territories and ruling that a slave was not a person who could sue in the courts. Prior to the Civil War, there was a much closer link between the Court's rulings and party policy than would be found later. This link reduced the Court's independence and thus the centrality of its role. During and after the Civil War, the Court also played a subordinate role—its usual posture with respect to war efforts—generally sustaining Lincoln's actions and his theory of the relation between Union and Confederacy.

The Court's general prestige in the late nineteenth century was great, but the Court did not play a positive leadership role. The Court undercut post–Civil War civil rights legislation, for example, in the *Civil Rights Cases*, and then, in *Plessy v. Ferguson*, sustained state legislation requiring Negroes to ride in separate railroad cars. At the same time, the justices were saying that the Fourteenth Amendment's Due Process Clause did not include the Bill of Rights' procedural protections for criminal defendants.[1] Later the Court's conservatism led it to uphold convictions of "subversives" in cases brought in the atmosphere of World War I.[2] The Court's importance in the late nineteenth and early twentieth century also resulted from its striking down most state attempts to regulate the economy even when Congress had not acted. This was the period when "substantive due process" was at its height—with the judges reading into the Constitution their views of what

*Citations to cases mentioned in the text may be found in the Table of Cases, p. 247.

was "reasonable" for the states to legislate. The Court also undid Congress's attempts to regulate the economy, defining the commerce and taxing powers narrowly and even saying that powers not expressly delegated to the federal government were reserved to the states—language like that of the Articles of Confederation.[3]

During the New Deal period, the Court became central politically as a result of its actions striking down much early New Deal legislation.[4] These decisions ran contrary to the political realignment reflected in President Roosevelt's election. The president made the Court an issue in his 1936 campaign and then in 1937 proposed that for every justice over the age of seventy who did not retire, an additional justice could be named. While Roosevelt said that this change was needed in the interest of judicial efficiency, it was clear that the president wished to use the new appointments to neutralize the justices opposed to his program. The "Court-packing" plan provoked a storm of disapproval even among Roosevelt's own supporters and was not adopted, but FDR really won the "war" against the Court when, in the "switch in time that saved nine," Justice Owen Roberts shifted his voting pattern, several conservative justices retired, and the Court sustained a wide variety of economic regulation measures.[5] After 1937 the Court was regularly to accept Congress's determinations as to the reach of the Commerce Clause, even as applied to racial discrimination.[6] (However, in 1976 the Court invalidated Congress's extention of the minimum wage to state employees, leading Justice William Brennan to raise fears that the Court was returning to its old ways.[7]) The Court, after 1937, was also to adopt a restrained stance toward state economic regulation, as long as such regulation did not directly interfere with national regulatory programs or impinge on civil rights.

The Court's post-1937 restraint in the economic field was not immediately accompanied by judicial protection for civil liberties. During World War II the Court again showed its deference to the president when it sustained the relocation of Japanese-Americans in the *Korematsu* case. Blacks began to win some support for their rights in voting, housing, and graduate education,[8] but the Vinson Court (1946–1953) had a generally conservative civil liberties record. Only with the Warren Court were minorities able to win victories they had not been able to obtain from reluctant legislatures and executives. Through its civil liberties "activism," most of it evident in decisions striking down state restrictions on individual rights, the Court again became extremely prominent and central to the nation's life. This was the only time in our nation's history when the Court has taken a noticeably pro–civil liberties stance.

> In relation to its predecessors, Warren's was the most activist court not only in America, but with a high degree of probability, also in world history, in

regard alike to the number and diversity of the civil libertarian causes that it sponsored, and the degree of favorable support that it gave to such causes.[9]

The Warren Court's first major step was the invalidation of "separate but equal" in education in *Brown v. Board of Education*, followed by 1956 and 1957 by protection for political dissidents caught up in governmental investigations of the Joseph McCarthy period. The Warren Court also required that both houses of each state legislature be apportioned on the basis of population; expanded free speech in the areas of obscenity and libel; invalidated prayers in public schools; and overturned convictions of those protesting against racial discrimination in transportation and places of public accommodations, also regularly sustaining Congress's actions to protect civil rights with respect to public accommodations, voting, and housing.[10] The justices also turned their attention to the poor, invalidating the poll tax and durational residence requirements for receiving welfare benefits.[11]

In what produced probably the greatest controversy, the Court adopted a series of broad rules protecting criminal defendants and suspects. Of these the most notable were *Mapp v. Ohio* (improperly seized evidence inadmissible at trial), *Gideon v. Wainwright* (right to counsel for indigents at trial for felonies and major misdemeanors), and *Miranda v. Arizona* (no confession admissible without suspect being warned of his rights). That the Court also approved "stop and frisk" practices, lifted restrictions on material which could be taken in a search, approved the use of informants, and supplied the basis for electronic surveillance with a warrant[12] received far less attention. So did the Court's avoidance of some problems, for example, the implementation of school desegregation for over a decade after the *Brown* decision and the basic constitutional question of the right of a private proprietor to refuse service on the basis of race—never reached despite all the sit-in convictions the Court reversed. Furthermore, the Court was criticized for having subordinated the interests of racial minorities to the concerns of whites and having too closely followed public opinion. The Court, it was said, had "struck down only the symbols of racism," leaving racist practices intact, and "had waltzed in time to the music of the white majority—one step forward, one step backward and sidestep, sidestep."[13]

The Warren Court's civil liberties record, reinforced by its easing access to the courts for those wishing to challenge government action, led to a considerable dependence on the federal courts. Groups increasingly turned initially to the courts for redress of their grievances, particularly against state and local officials, instead of using those courts as a last resort. This gave pause to the Court's observers, including some who supported its decisions. They wondered whether reliance on the courts might lead to atrophy of the legislative process, central to a democratic political system. As Judge J. Skelly Wright wrote:

It would be a very serious matter for the people to come to rely on the Court as the most important force for change in our society. That the legislature might come to enact laws without adequately considering their constitutionality, leaving it to the Court to correct its mistakes, would be very unfortunate; but if the people and the legislatures were to come to expect the Court to do their work for them, it would be far worse.[14]

The type of concern voiced by Judge Wright substantially abated in the 1970s as a result of the Burger Court's actions—its adoption of a conservative posture on criminal procedure issues, its general unwillingness to go beyond the Warren Court's position in other areas, and, perhaps most important, its enunciation of decisions making it considerably more difficult to bring federal court suits challenging the alleged deprivation of civil liberties. This change in position began to lead people to turn for resolution of their grievances to state courts where such courts were liberal and, in general, to political rather than judicial processes.[15] The balance of influence between the courts and the elected branches of government thus moved toward its pre–Warren Court position. The continuing validity of some Warren Court precedents and the Burger Court's own affirmative civil liberties decisions, including those on employment discrimination and women's rights, continue to give the Court a greater prominence concerning civil liberties than it had before Earl Warren became Chief Justice. However, despite its continuing prominence, it is clear that the Court will not be as predominant or central in our political life as it was during the 1950s and 1960s.

THE COURT IN TRANSITION

The shift—both real and perceived—between the Warren Court and the Burger Court has focused attention on transitional periods in the Supreme Court's history. Perhaps all courts are courts in transition. As they change personnel, their policy direction changes and, as noted earlier, so does the relationship of the Court to public sentiment—at times moving closer to public opinion, at times appearing to lead that opinion. Such changes also involve shifts in the predominant role the courts assume, whether as restrained discoverers of the law or more actively as developers or makers of policy. During a transitional period, several years may elapse before a pattern is clear. Moreover, different patterns may characterize different policy areas. The "new" Court may extend earlier rulings; maintain, consolidate, or clarify those rulings; or curtail or erode the precedents of previous years— often with substantial cumulative effect. Direct reversal of precedent may occur, but usually only after several years have elapsed. In the words of one observer, "After the Marshall era, a period of great constitutional creativity, the Court experienced a period characterized by limitation and modification but not by major departure from established precedent."[16] In some ways, that is the position the Burger Court held vis-à-vis the Warren Court.

The transition to the Roosevelt Court, which began prior to a change in personnel and then was reinforced by new justices like Hugo Black and William O. Douglas, did not take place until FDR's second term, when his first opportunities to nominate justices occurred. The transition from the Vinson Court to the Warren Court was immediately visible in the 1954 ruling in *Brown v. Board of Education* on school desegregation. However, a few terms passed before the Court "succeeded impressively in freeing itself from the self-doubts that deterred constitutional development during the 1940–1953 period."[17] For example, after its liberal internal security decisions of 1956 and 1957, the Court retreated in the face of a concerted congressional attack on its jurisdiction, with the Court supporting civil liberties claims less than half the time in 1958–1960. The Court's support of such claims at rates of close to 80 percent came only when Justice Felix Frankfurter retired in 1962 and was replaced by Justice Arthur Goldberg, the Court's fifth reliable civil liberties vote. Even with Goldberg present, in 1964 there was a noticeable drop in the percentage of civil liberties claims supported. However, the Warren Court did "finish strong," providing an extremely clear contrast with what was to follow.

Unlike FDR, Richard Nixon had immediate opportunities to select justices. Nixon was able to select a Chief Justice and another justice in his first year in office—although the Senate's defeat of two nominations delayed the second appointment—and two more two years later. Having made the Court, and particularly its criminal procedure rulings, a subject of campaign controversy, Nixon gave particular attention to his nominees' ideology. Initially, Chief Justice Burger produced no great across-the-board policy change, although there was some limited retreat and some areas, most notably criminal procedure, saw more change than others. Although the Court limited expansion of Warren Court doctrine, the basic picture for several years was one of marginal change and a generally unsettled pattern, with the 1974 Term leaving one observer with a picture of a "reluctant Court" that

> tries to make potentially important cases stand for as little as possible; a Court whose theme song is the refrain "we *only* decide"; a Court that loves to decide issues "in these particular circumstances," that performs contortions to avoid announcing new principles even when new principles are inescapably needed, and that pretends not to be announcing them even while it is announcing them ...[18]

At first, precedent was not overruled. Instead the Court seemed to use an approach of "whittling away of precedents . . . to the point that overruling them may be unnecessary."[19] However, in the 1975 Term, the Court did overrule a 1968 Warren Court precedent on picketing of a store in a shopping center; one of its own decisions concerning the handing out of leaflets on a military base; the 1968 decision upholding Congress's extension of the minimum wage law to state employees; and a 1949 ruling allowing state regulation of peaceful activity by workers not subject to the National Labor Relations Act; and, in addition, by allowing seizure of business records, the Court abandoned the ninety-year-old doctrine of *Boyd v. United States* that private papers could not be subpoenaed because they contain incriminatory statements.[20] Such action shows that although an allegedly self-restrained or "strict constructionist" Court is thought to be one which will follow precedent, the justices' values—here, particularly their conservatism on civil liberties—will influence the decisions they reach.

The Burger Court, even before it consolidated its position and reversed precedents directly, showed through both doctrine and results considerable withdrawal from and undercutting of Warren Court policies affecting the entire range of civil liberties problems but particularly noticeable in the criminal procedure and free speech areas. In Burger's first term, support for civil liberties claims fell to 55 percent, and by the 1972 Term it was only 43.5 percent, the lowest since 1957.[21] Indicative of the change is that in the 1974 Term, most unanimous civil liberties decisions were in favor of the civil liberties claim, but when the Court was divided, the claim was rejected 29 out of 46 times. Schubert shows that if the late Warren Court ranked first on political liberalism (civil liberties) for the period from the mid-1930s to 1972, the Nixon Court ranked behind the middle Warren Court and the Roosevelt Court, putting it at "about the point where it was when [Nixon] became vice-president and Earl Warren was in his first term or two."[22] Even greater change occurred with respect to economic policy, with the Nixon Court ranking just above the Hughes Court of the 1930s which FDR had battled.

Overall, the Burger Court gave greater weight to claims by the government—at both the national and state levels—and less to claims by the individual citizen than had the Warren Court. An example of its deference to the executive branch occurred in *Kleindienst v. Mandel* when the attorney general would not grant a visa to a foreign Marxist theoretician invited to the United States to give lectures. Despite the dissenters' complaint that the reason offered by the attorney general was a "sham" and that the Court was giving "unprecedent deference to the Executive," and despite the First Amendment interests of the potential audience, the Court refused to scrutinize the executive's reason for denying the visa. The new Court also subjected challenged state laws to less strict tests than the Warren Court had used and returned to the case-by-case approach in the criminal procedure area. The Court was also less willing to serve as a forum for resolving complaints against the government. When a policeman complained that he had been improperly discharged without a hearing, the Court said, "The federal court is not the appropriate forum in which to review the multitude of personnel decisions that are made daily by public agencies. . . . The United States Constitution cannot feasibly be construed to require federal judicial review for every such error."[23]

Despite this basic picture and a definite retreat from the Warren Court's reapportionment rulings—greater flexibility was allowed for state legislative districts than for congressional districts, and the Court refused to set aside multimember legislative districts which provided less opportunity for blacks to be elected as long as the blacks had access to the political process[24] —there were areas where the Court moved forward with Warren Court doctrines and other areas not considered by the Warren Court where new ground was broken. Lower court judges' authority to order broad remedies for school segregation, including busing, was upheld; schemes for evading desegregation were invalidated; and private schools were told they could not discriminate on the basis of race.[25] However, in a ruling seen by some as the "end of the road" for desegregation, the Court set aside a plan for busing across school district lines where it could not be shown that the other districts had themselves discriminated; a metropolitan-area-wide remedy for public housing discrimination was, however, later approved.[26] Other "negative" rulings included those that a private club did not have to serve a black guest merely because it had a state liquor license—thus limiting the scope of the Fourteenth Amendment—and that a city (Jackson, Mississippi) could close all its swimming pools for economic reasons, even though the dissenters said it was being done to avoid desegregation. However, the Court took a strong stand against discrimination in employment, reinforcing what Congress had done. The justices said non-job-related employment tests with a discriminatory effect could not be used even though no discrimination was intended, and courts were allowed to grant back pay and seniority rights to remedy discrimination.[27]

In the area of women's rights, the Court invalidated an automatic preference for men as executors of estates; statutes making it easier for men than women to obtain dependents' benefits; automatic exemptions of women from jury duty; forced maternity leave requirements for school teachers; and laws interfering with the right to obtain an abortion, particularly during the first trimester of pregnancy.[28] However, the exclusion of pregnancy from disability insurance coverage was upheld, as was a property tax exemption for widows but not widowers.[29] The Court's attitude toward welfare recipients was, however, quite unfavorable. The Court did say in 1970 that welfare benefits could not be discontinued without a hearing, but refused to require states to increase welfare benefits to reflect cost-of-living increases, sustained "maximum grant" provisions, and upheld removal of benefits when a welfare recipient would not allow a social worker in her home without a warrant.[30]

In First Amendment cases, the Court continued—and advanced—a strong position on separation of church and state while its free speech posture was mixed. The justices permitted construction financing and broader aid for secular functions at the college level and free textbooks for parochial schools but outlawed virtually every other form of aid to parochial schools at the elementary and secondary levels—tuition grants, payments for services or for maintenance of buildings, and salary supplements.[31] The Court also upheld the property tax exemption for churches and sustained the right of Amish parents to limit their children's schooling.[32] With respect to free speech, the Court refused to allow the government an injunction against publication of the "Pentagon Papers" and struck down "gag orders" on the press in criminal trials. However, the First Amendment was held to allow grand juries to question reporters about their confidential sources, and free speech by military officers and on military bases was restricted.[33] The Court distinctly moved away from the Warren Court's position on obscenity, in *Miller v. California* rewriting the definition of the concept to give greater scope to local community values.

The Burger Court's greatest retrenchment from Warren Court doctrine occurred with respect to criminal procedure, particularly searches and confessions. The rule excluding improperly seized evidence was held inapplicable to grand jury proceedings or civil proceedings and the Court limited federal court review of claims about improper searches.[34] Although some Border Patrol car searches were invalidated, searching people and cars was made easier and the ability to seize a person's records—bank records in one's own possession—was facilitated.[35] The Court also limited the immunity necessary before a person could be compelled to testify.[36] *Miranda* was not overturned but its effect was weakened; questioning of a suspect could be resumed several hours after it had initially been terminated; confessions obtained without warnings could be used to impeach a witness's testimony at trial; and grand jury testimony and tax investigations could be carried out without the *Miranda* warnings of rights.[37]

While the Court extended right to counsel at trial to all cases where a person was to be jailed, the justices allowed states to recover money spent in providing free counsel, limited right to counsel at line-up and on appeal, and refused to make counsel automatically available for those whose probation or parole was being revoked or to provide counsel in prison proceedings for rule violations although procedural protections were established in those situations.[38] Of particular importance was the Court's legitimizing and encouragement of plea bargaining and its strict limitations on challenges to a conviction once a defendant had entered a counseled guilty plea.[39]

This brief partial summary of the Burger Court's rulings indicates a Court which, while it reinforced or advanced some Warren Court doctrine and developed rights in previously unexplored areas, was perhaps more important through its action eroding major Warren Court rulings and limiting access to the courts, because through the latter actions it moved closer to the more conservative public opinion on civil liberties which had developed in the late 1960s and responded to criticism of judicial "activism." While substantially conservative on search and seizure and Fifth Amendment policy, the Court has been liberal in other areas, like some aspects of racial and sexual equality; this is certainly true if our comparison is with the pre-Warren Court era. Although the Court has begun to consolidate its decisions, there is still a high "surprise level" in its actions. The appointments to the Court made by President Carter will help determine whether the Court continues on its present path or moves in a more liberal direction—or at least one which makes it easier for those challenging disliked laws and statutes to get into court to try to "make their case."

FINDER OR MAKER OF THE LAW?

For many years the principal belief about the Supreme Court was that it did not make, but only found, the law. The natural law concept that there was an external and immutable truth which could be found through "right reason" took the form that judges found preestablished law in statutes, regulations, and particularly the Constitution, applying in specific cases the intentions of those who wrote those documents. Reverence for the Supreme Court, which was "above politics," reinforced this view: the Constitution had become our bible and the Supreme Court justices our high priests, with the public transferring "our sense of the definitive and timeless character of the Constitution to the judges who expound it."[40] The belief that the judges found rather than made the law has had considerable force and has been enunciated even by the judges themselves. For example, the judges' limited role was stated by Justice Owen Roberts, in the *Butler* case, to be only that of placing a challenged statute alongside the Constitution: if they matched, the law was constitutional; if they did not, the law was invalid. (Roberts's view has come to be known as the "slot machine theory of jurisprudence.") The belief that judges only found the law was perpetuated even by some who realized it to be a myth, because they thought the myth gave legitimacy to the Court's decisions.

Despite the myth, the Supreme Court does make policy. The Judges' Bill of 1925, which gave the Court its present power to select the cases it would hear, helped make clear the Court's policy-making role. A court required to decide all cases brought to it—and thus deciding a smaller proportion of controversial cases—would be more easily seen as a "regular" law court. Yet it is hard to say that a court, particularly the highest court in the land, which chooses cases, concentrating on those affecting broad classes of people, is not making policy. This is particularly true when those cases often produce division within the Court. The myth may not have been difficult to believe when the Court's rulings conformed closely to the nation's predominant political ideology. However, the conflict between Roosevelt and the Court, when the Court's position was not shared by much of the population, helped considerably to alter the view that the Court found law and to produce a realization that, as Oliver Wendell Holmes put it, Supreme Court decisions were not brought by constitutional storks—that judges exercise discretion in interpreting the Constitution's ambiguous language. Chief Justice Charles Evans Hughes's statement that the law is what the judges say it is further confirmed the new idea. What the conflict between FDR and the Court had started, the Warren Court's actions put beyond doubt—at least as to what the Court actually did, even if not as to what it *should* do. Resistance to and attacks on the Court's school desegregation, internal security, reapportion-ment, school prayer, and criminal procedure rulings made clear that the Court's majority both made policy and often held views that differed from those held by important segments of the politically active public as to basic matters in the Bill of Rights.

Despite such recognition of the Court's making of policy, it must also be kept in mind that there are differences from one policy area to the next in the degree to which the Court makes policy. Sometimes individual decisions have great policy effects; at other times it is through the cumulative effect of the decisions that policy develops. The Court may be a "policy leader" or may avoid policy making; it may accept Congress's statement of the need for a statute or an administrative agency's factual determinations or may resolve such matters for itself. Such variation "might seem rather strange in an isolated and insulated court administering The Law by processes of rigorous legal logic," but is not unusual "in a political agency faced with a wide range of problems, each entailing a different constellation of political forces."[41]

Virtually all who follow the Court's work now admit that the Court does make policy and cannot avoid doing so. Some of the judges have even openly acknowledged that this is the case. Justice Byron White, dissenting in *Miranda v. Arizona,* argued that the majority's ruling, while not exceed-ing the Court's powers, served to "underscore the obvious—that the Court has not discovered or found the law in making today's decision, nor has it derived it from irrefutable sources." Instead it had made new law "in much the same way that it has done in the course of interpreting other great clauses of the Constitution. This is what the Court historically has done. Indeed, *it is what it must do* and will continue to do until and unless there

is some fundamental change in the constitutional distribution of governmental powers."[42]

Yet from time to time judges try to say that they do not engage in policy
making based on their personal values—that they act differently as judges
from the way they would act in other situations. As Justice Frankfurter put
it in 1943 while dissenting when the Court invalidated the compulsory flag
salute for school children:

> As judges we are neither Jew nor Gentile, neither Catholic nor agnostic. As a
> member of this Court I am not justified in writing my private notions of policy
> into the Constititon. . . . The duty of a judge who must decide which of two
> claims before the Court shall prevail . . . is not that of the ordinary person . . .[43]

Justice Blackmun, dissenting in 1972 when the majority invalidated the
death penalty as then applied, said the punishment "violates childhood's
training and life's experience" and that as a legislator he would sponsor
legislation to repeal the penalty and as a governor he would use executive
clemency to prevent people from being executed. But, he said, "There—on
the Legislative Branch of the State or Federal Government, and secondarily,
on the Executive Branch—is where the authority and responsibility for this
kind of action lies."[44] Similarly, Chief Justice Burger, voting to sustain an
order of the Federal Communications Commission, said he was unsure that
the FCC had made the right decision, but that as a justice of the Supreme
Court, he could not "resolve this issue as perhaps I would were I a member
of the . . . Commission."[45]

Such statements, particularly Frankfurter's, are unrealistic. Indeed, for
those interested in the Court's results, they are irrelevant because, for the
result-oriented, it is the votes—to sustain the flag salute or affirm use of the
death penalty—which really count. And such statements have often been
used in aid of conservative decisions, for example, during the Burger Court
period. Certainly a judge's views of what a judge should do—his role con-
ception of what it means to be a judge—can have some effect on his actions.
Yet a judge cannot fully put behind him, or put to one side, all past experience and personal values, particularly when deciding highly controversial
cases. These statements, whether or not they are rationalizations for policy
positions the judges would have reached in any event, are important none-
theless. Coupled with and reinforced by beliefs still held by many that the
Court finds the law and that it should act in special, "nonpolitical" ways—
that it is not just "any old political actor"—the statements indicate that the
Supreme Court must continue to *act* like a court. As Dahl asserts, "If the
Court were assumed to be a 'political' institution, no particular problems
would arise, for it would be taken for granted that the members of the Court
would resolve questions of fact and value by introducing assumptions

derived from their own predispositions or those of influential clienteles and constitutencies."[46] However, the Court—like any policy maker—must act within the context of the beliefs people hold about it, beliefs which constrain its actions.

Neutral Principles

Among the expectations which people—those who see the Court as finding law and even many who feel it makes policy—have of the Court is that its decisions should be *principled*. Pragmatic compromises are for legislatures and executives; courts must stand for principle, deliberateness, the use of rationality and logic, and detachment from the turmoil and passion of political conflict. In Bickel's words:

> The root idea is that the process is justified only if it injects into representative government something that is not already there; and that is principle, standards of action that derive their worth from a long view of society's spiritual as well as material needs and that command adherence whether or not the immediate outcome is expedient or agreeable.[47]

In the most noted statement, Herbert Wechsler said judicial determinations should rest on "reasons that in their generality and their neutrality transcend any immediate result that is involved." Criteria for decisions had to be exercises of reason, "not merely . . . an act of willfulness or will."[48] But Wechsler had demanded too much even for some advocates of the Court's "passive virtues." Bickel argued that if the Court were to have to rest its decisions "only on principles that will be capable of application across the board and without compromise, in all relevant cases in the foreseeable future," with any flexibility built into the principle itself "in equally principled fashion," there would be few such principles; few cases would be decided and the Court would thus fail to meet the public's demands that it resolve issues. (Holmes once remarked that it is better that some cases be decided than that they be decided right.) It should be enough, Bickel said, to have "an intellectually coherent statement of the reason for a result which in like cases will produce a like result, whether or not it is immediately agreeable or expedient," or a "principled process of enunciating and applying certain enduring values of our society," with the values having "general significance and even-handed application."[49] However, new principles might arise when there was a conflict between values, for example, between the right of people to associate—which Wechsler thought must have been the basis of the decision in the school desegregation cases, the opinion in which he criticized—and the right of others not to associate, or the right of free speech and the right to a fair trial, faced by the Court more recently.[50]

A serious problem with the demand for "neutral principles" is that, while initially appealing, they might turn out to be rigid and thus prevent the Court from reaching publicly acceptable results: the public may want its principles but it also wants decisions with which it can live. For example, the statement of Justice Harlan, dissenting in *Plessy v. Ferguson,* that "the Constitution is colorblind," sounds marvelously unbiased but, taken literally, would prevent most programs to remedy results of past racial discrimination. Thus the Court is faced with the dilemma of having simultaneously to reach toward principle and reasoned judgment as the basis for its actions while maintaining some flexibility in order to adjust to changing social conditions.

Precedent and Incrementalism

An even stronger expectation, particularly within the legal community, is that in order to provide consistency and uniformity—and thus fairness— decisions be based on precedent, the doctrine of stare decisis (literally, to stand on the decision) which is at the heart of the Anglo-American legal tradition. The expectation that courts base their decisions on precedent, which involves taking the present case, finding past cases which are similar, and then applying the rules from those past cases, creates a number of problems. How does one know which cases are "similar"? On what competing precedents should one rely? How should legal doctrine of the relevant cases be interpreted? And, at the level of the Supreme Court, should precedent be followed as closely as lower courts are expected to follow it? The Court's place as the nation's highest and last court has led it to feel *less* bound by precedent than the lower courts, particularly where interpretation of the Constitution is involved because changes in constitutional rulings can come about only through the very difficult process of amending the Constitution itself.[51]

Any time the Supreme Court majority overrules a precedent, criticism both from dissenting justices and from outside the Court is likely. This is true whether the precedent is new or old; if old, the Court is said to be interfering with the "settled ways" of the law; if new, the Court is not allowing the law to develop properly. Particularly where departure from recently established precedent has not been explained well by the justices, such criticism is often more than merely a "cover" for opposition to the result the Court has reached. An example is provided by the Burger Court's recent rulings on leafleting in shopping centers. In 1972 the Court had said that antiwar leafleting could be banned in a privately owned shopping center because the protest was unrelated to the shopping center or the stores there. The Court did *not* overrule a 1968 case allowing picketing of a particular store in such a shopping center but instead said the situations in the two cases were different. Then in 1976 the Court banned picketing of a shoestore in a shopping center by workers whose basic dispute was with the store's warehouse. Here the Court explicitly overruled the 1968 case but in doing so said the overruling really had taken place in 1972 because the first (1968) case had not "survived" the second (1972) one. The inadequacy of this explanation, so obviously at variance with the majority's own 1972 statements, prompted dissenting Justice Marshall to complain that the first case had "been laid to rest without ever having been accorded a proper burial."[52] Similarly, the Court's overturning of Congress's extension of the minimum wage law to state and local government employees, in which it overruled another 1968 decision upholding a closely related extension of the law, led Justice Brennan to argue that the Court had "discarded roughshod" a whole series of decisions.[53]

If explicit overruling of precedents produces criticism, the justices may choose to achieve their desired results by "distinguishing" precedents, that is, by limiting the application of those precedents by saying new cases contain facts not identical to those in the older cases. The 1972 antiwar leafleting case just noted provides an example. The Burger Court also used the technique in its erosion of Warren Court criminal procedure precedents. Where the Warren Court had restricted the extent of a "stop and frisk" search of a person not under arrest, the Burger Court, stressing the difference between a more "casual" *stop* and a more extended *arrest*, allowed much more extensive searches of someone who had been arrested.[54] The Warren Court's extension of right to counsel to police-conducted line-ups was limited to postindictment line-ups by emphasizing facts the Warren Court had not stressed, and the Warren Court rule was also said not to apply when a witness was examining a display of pictures rather than viewing a line-up of people.[55] Erosion of *Mapp v. Ohio* and *Miranda* occurred in much the same way. Yet cutting away at precedents by distinguishing them is not without problems, as critics, including other justices, demand that the Court face up to past rules whose thrust is being ignored.[56]

Despite the Court's overturning and distinguishing of precedent, the expectation that precedent will be used has a considerable effect on at least some of the justices, even when they disagree with the past cases. For example, the Court's 1976 ruling invalidating discrimination on the basis of race in admission to private schools involved an application of its 1968 *Jones v. Mayer* decision applying post-Civil War antidiscrimination statutes to private housing sales. Justice Powell went along with the majority in 1976 because the Court had considered application of the civil rights statutes to private acts "maturely and recently," while Justice Stevens indicated he would vote differently "were we writing on a clean slate." However, Stevens said that "the interest in stability and orderly development of the law" and the fact that the precedent "accords with the prevailing sense of justice today" had "greater force" than the argument that the earlier case was wrongly decided.[57]

Precedent, coupled with the expectation that previous cases will be interpreted in certain generally accepted ways, has exerted a "brake" on the Court's actions, causing it to move gradually. As Walter Murphy has remarked, "When the Court reverses itself or makes new law out of whole cloth—reveals its policy-making role for all to see—the holy rite of judges consulting a higher law loses some of its mysterious power."[58] The Court has thus discerned the need to make its decisional output appear incremental, to stress continuity even where great changes are taking place. Too many sharp deviations and departures, too many unpredictable results disturb the general public and particularly those "opinion leaders" who watch the Court most closely.

Gradual movement might well characterize the Court's decisions even without precedent, as a result of deciding narrow questions before large ones or making summary disposition of appeals instead of rulings with full opinion. Most policy makers do not make radical departures from past policy but operate incrementally, developing policy a bit at a time and making changes "at the margin." As Shapiro has argued, "The theory of incrementalism may explain, or at least describe, the phenomena of stability and gradual change in law just as well as or better than stare decisis. . . ."[59] The school prayer decisions, where the Court first struck down a state-written prayer in *Engel v. Vitale* before eliminating all recitation of school prayer and Bible reading in *Abington School District v. Schempp*, provides an example of incremental action. So does reapportionment, where the Court first demanded equally populated districts for the U.S. House of Representatives (*Wesberry v. Sanders*) and invalidated the Georgia "county unit" system (*Gray v. Sanders*) before ruling, in *Reynolds v. Sims*, that both houses of all state legislatures had to be apportioned on the basis of population. In the area of criminal procedure,

The Court would typically approach a new issue warily, issuing first a narrowly limited decision which contained a hint of the result that might finally be required. A few criminal lawyers would get that hint, cases would develop, and lower court decisions would result. These decisions would usually not be consistent with each other, which could be a boon to the Supreme Court; it could deny petitions for certiorari in all of these appeals, leaving the lower court rulings in effect without indicating whether it approved or disapproved of the results. Over the years experience would develop in the lower courts as to the best way to proceed with the problem, and any warning signs would be detected before the Supreme Court had to take its position.[60]

While policy may be developed incrementally, it may also be developed in "big pieces" and *applied* incrementally, as successive litigation brings the various parts and pieces of a problem to the Court. Thus the Court first provided a broad definition for obscenity in its 1957 *Roth* decision, gradually adding elements to the definition in later cases.[61] After the Court said in the *Brown v. Board of Education* that "separate but equal" was not to apply to education, the Court applied this rule to other public facilities (golf courses, swimming pools, and the like) through a series of unsigned (per curiam, "through the Court") opinions shortly afterward. These examples also show the incremental development of what *seems* to be a major policy change, as the important elements of the *Roth* obscenity definition were derived from earlier lower court rulings, and *Brown* built on *Sweatt v. Painter*, invalidating segregation in graduate education because "intangible" factors were not equal.

ACTIVISM AND SELF-RESTRAINT

Neutral principles and precedent are not enough to restrain the Court. One thus finds continuous controversy over whether the Court has been sufficiently "self-restrained" or too "activist." Justice Harlan Fiske Stone once said that the only restraint on the Court was its own sense of self-restraint. The question then becomes whether and when it should exercise that restraint, whether, for example, as Justice Frankfurter—during his tenure the foremost advocate of judicial self-restraint—argued, the Court should stay out of the reapportionment issue because it involved "political entanglements" and the "clash of political settlements"[62] or whether the Court should take an active role in protecting individual rights.

The Court is said to follow some basic "rules" of self-restraint perhaps spelled out by Justice Louis Brandeis.[63] The Court will not rule on challenged legislation in a nonadversary proceeding—that is, where the parties are not in conflict with each other—nor will the Court decide a complaint made by someone not injured by the statutes or who has benefited from the challenged statutes. A constitutional issue will not be decided if other bases for decision, such as statutory construction, are available; in short, the Court will not anticipate a constitutional question unnecessarily nor will it formulate a rule broader than required by the facts of the case.

Considerable disagreement has occurred over whether the Court has followed these "rules," particularly as to whether it has gone beyond the facts of the case. The Warren Court's broad criminal procedure rules have drawn particular criticism from those who thought cases should be decided on the basis of the "totality of the circumstances"—including defendant's age, prior experience with the law, and seriousness of the offense. Generally the breadth of the Court's doctrine is related to judges' goals; when faced with broad rules favoring law enforcement, liberals argued for the "totality" test. Similarly, self-restraint and activism are generally related to the judges' values. At times, deferring to legislative action can help achieve policy goals, while others can be obtained only by using rigid tests to overturn statutes. Fifty to seventy years ago, self-restraint meant not disturbing national and state legislation regulating the economy and produced "liberal" results, while "activism"—striking down statutes as unreasonable interference with "freedom of contract"—assisted the business community. In the last forty years, and particularly the last twenty, a "hands off" approach preserved statutes infringing on individual freedoms, while "activism" has generally protected civil liberties. Richard Nixon, who pledged to appoint "strict constructionist" judges who would not encroach on areas belonging to Congress and the president, criticized judges for having "gone too far in assuming unto themselves a mandate which is not there, and that is, to put their social and economic ideas into their decisions."[64] Yet his remarks about Supreme Court decisions which set free "patently guilty individuals on the basis of legal technicalities" meant he did not want self-restrained judges when activism could aid the "peace forces" rather than the "criminal forces."

Until the Warren Court, the Supreme Court had seldom given protection to minorities. After 1937 the Court approached civil liberties issues in a less restrained fashion than earlier, although justices have differed over whether to be more activist in some areas than others. Where statutes *protect* rights, the Court has usually been quite willing to defer to the legislative judgment, a position which could be seen in the Warren Court's approval of civil rights laws. It could also be seen in a different way when Justices Powell and Blackmun made what amounted to a request that the Equal Rights Amendment (ERA) be disposed of so the Court would have more guidance in deciding women's rights cases instead of having to be more actively involved through interpretation of the more obscure language of the Fourteenth Amendment's Equal Protection Clause.

The decreased deference to laws impinging on civil liberties and civil rights can be dated from 1938, when Justice Stone suggested that legislation dealing with the exercise of liberties, particularly if it affected minorities' access to the political process, would not be given the same presumption of validity accorded to legislation regulating the economy.[65] Those who adopted the "preferred position" doctrine, that free speech was more important than other rights, have been particularly critical of laws touching on rights in that area while being more restrained about other statutes. Others, however, have felt that statutes in all areas of policy should be treated with the same degree of deference, with the Court playing only a limited (self-restrained) reviewing role.[66] In part as a result of these—and other—conflicting views, the Court has not been consistent in the tests it has applied to statutes challenged as violating liberties. The most restrained position has been that a statute need bear only a reasonable relation to the government's interests. The most activist is that it must pass "strict scrutiny"—a test used when a "suspect category" like race or alien status is in the statute. Yet justices vary over time in the tests they use and, as measured by results, may appear to be using one under the guise of another.

The argument over "activism" or self-restraint concerning civil liberties —seen most clearly through comparison of the Warren Court and the Burger Court—is extremely important. Regardless of the Court's position, the elements of the Bill of Rights are rights only in theory. Because the pressures in the "real world" against civil liberties are substantial, they seldom receive sufficient protection and are continually attacked by private groups and by the government, as FBI surveillance activities illustrate. If the Court does not stress constitutional rights—with the positive effect on popular views such judicial action can have—no other institution in our political system is likely to do so. True, by itself the Court cannot guarantee protection for our liberties. If the Court engages in self-restraint, deferring to government officials on civil liberties issues rather than vigorously asserting the need to enforce our rights, people are encouraged to believe that civil liberties need not be taken seriously.

Yet adoption of a strongly activist position with respect to individual freedoms is not necessarily realistic for the Court as a political actor. Despite approval from some segments of the public, the Warren Court's decisions were frequently attacked and often disobeyed. Indeed the Court almost lost some of its jurisdiction and perhaps would have suffered severe damage had it not avoided some issues and retreated on other occasions. The absence of attacks on the Burger Court, a result of the greater congruence of its decisions with the nation's civil liberties orientation, suggests the virtues of judicial self-restraint. Those virtues could also be seen after 1937 when the Court changed its posture with respect to economic regulation to one of self-restraint. The unanswered question is whether and how the Court can preserve its independence without simply becoming a sterile institution following public opinion and having little effect.

THE COURT AND STRATEGY

A question like the one just posed leads to a consideration of judicial strategies. Changes in the Court's positions—from liberal to conservative, from activism to self-restraint—are not accidental. In large part they result from the arrival of new justices—often chosen by presidents for their ideological proclivities—and the types of cases which come to the Court, both "external" forces, and attitudinal changes over time by individual judges, an "internal" factor. The judges contribute another important internal element through their strategies. Those who believe in the myth that judges find the law and those who seek to have the justices apply "neutral principles" find unseemly any talk of the Court engaging in strategy. They want issues confronted directly, regardless of the cost, and do not wish justice for the parties sacrificed for other concerns. They are disturbed by such actions as the Court's refusal in the 1950s to decide a challenge to a cemetery's refusal to bury a part-Indian or to rule on antimiscegenation (mixed marriage) laws —even when a black woman went to prison for marrying a white[67]—or, more recently, soldiers' challenges to the constitutionality of the Vietnam War, despite the increasing death rate during that conflict.[68] Yet because the Court as a policy maker is a political body, acting in an environment which is not necessarily supportive and may become quite threatening, by definition it must act strategically. As Ulmer has commented: "Support for civil liberties, as provided by the Supreme Court, is a commodity for which buyers' markets alternate with sellers' markets in some kind of cyclical pattern. Courts of law can be expected to 'read' their markets as well (hopefully) as any other producer of consumables."[69] If it is to survive and be able to achieve some of its goals, the Court may have to avoid some controversies. To be avoided where possible are "self-inflicted wounds" like the *Dred Scott* decision legitimating slavery and, some would add, the Warren Court's criminal procedure rulings. A justice's 1954 comment, "One bombshell at a time is enough," made when the Court turned away a miscegenation case, was a recognition of the need to concentrate on achieving school desegregation—affecting a large number of people—rather than damage that effort by angering the South with a decision on the mixed marriage laws, affecting relatively few individuals.[70]

That the Court does take actions based on strategy is clear from the evidence provided by the justices themselves. For example, the decision in *Brown v. Board of Education* was delayed to get a vote which was unanimous or more nearly so.[71] When Justice Frankfurter was assigned to write the Court's opinion striking down the "white primary," the Chief Justice was told that, because the ruling was likely to be received negatively in the South, it would be wiser if the opinion were *not* written by a Jewish immigrant who had taught at Harvard Law School—and the case was reassigned to Justice Stanley Reed, a Kentuckian.[72] Similarly Justice Byrnes, also a southerner, was assigned the opinion in a case holding that a Georgia statute violated federal prohibitions against peonage, in order to give it more force.[73]

The justices' demonstrated awareness of their political environment makes it quite likely that, acting on the basis of strategic considerations, they take that environment into account even when they claim not to do so. Justice Powell, in upholding school financing through the property tax, said,

"Practical considerations, of course, play no role in the adjudication of the constitutional issues presented," but commented that to rule otherwise "would occasion . . . an unprecedented upheaval in public education."[74] In the field of criminal procedure, the justices' taking the effect of their rulings into account has been most obvious. For example, Justice Stewart justified a new restriction on admissibility of seized evidence by pointing out that the FBI and the federal courts had remained effective although operating under the exclusionary rule for many years; in the 1976 death penalty decisions, several justices said that the fact that thirty-five states had reenacted death penalty legislation after statutes had been struck down in 1972 was an indication that the penalty was not "cruel and unusual punishment"; and Justice Clark asserted that to apply the exclusionary rule of *Mapp v. Ohio* to past cases "would tax the administration of justice to the utmost."[75] One of several criteria the Court regularly used in deciding whether to apply its criminal procedure rules to past convictions was the effect such application would have upon the administration of justice. Indeed, the Court's approach to the retroactivity problem has been said to have been guided largely by strategic considerations such as placating those opposed to the Court's new rules.[76] It is clear that the Court has acted strategically in responding to threats at other times as well, either through changing its policy position on economic regulation after FDR's attempt to pack the Court and on internal security after congressional efforts in the 1950s to limit its jurisdiction, by refusing to decide more cases on a controversial issue, or by deciding cases with larger majorities.[77]

To say the Court acts strategically does not mean that the justices have a complete blueprint to guide their actions. It is more likely that individual justices act strategically than that there is a "strategy of the Court." Although there may be a general sense of direction and the justices appear to share the goal of preservation of the Court as an institution, the Court is perhaps best thought of as a "mixed-motive" or "mixed-strategy" group. Even if all the justices were concerned about strategy at the same time, they might focus on different aspects of a case or wish to pursue different strategies. Moreover, although a "policy-oriented judge" is "aware of the impact which judicial decisions can have on public policy, realizes the leeway for discretion which his office permits, and is willing to take advantage of this power and leeway to further particular policy aims," not all justices may fit this description: "Probably relatively few justices have had a systematic jurisprudence; more but probably still relatively few have been so intensely committed to particular policy goals as to establish rigid priorities of action that dominated their entire lives; probably few have been able to act only rationally in seeking to achieve their aims."[78]

There are also numerous constraints on a judge's ability to accomplish any strategic aims. Each judge "has only a finite supply of time, energy, research assistance, and personal influence."[79] Strategic considerations often compete and there are standard operating procedures (SOPs) and "bureaucratic routines" which courts, like any other organization, follow. The justices respond to self-imposed deadlines like the one that all cases argued within a term are to be decided by the end of that term, as well as by rules for handling caseload, including technical ones on timely filing and judicial doctrines on the standing of a party to raise an issue. These rules can be ignored when it is felt necessary, but if the Court is to function effectively, that cannot be done often. Other constraints on the Court's exercise of strategy include the judges' own positions, interpersonal relations within the Court,

> the state of the Court's prestige and professional reputation; . . . the status of public opinion in general and the relative strengths and skills of the groups most apt to press for threatening or supportive action; . . . the degree of congressional commitment to the statute's policy; the attitude of the President and other executive officials toward the statute as well as the state of their current relations with Congress; . . . and the Court's control over its own bureaucracy.[80]

These constraints affect both the Court's ability to have an effect on the political system ("external" strategy) and the persuasion and bargaining among the judges while the Court has a case under advisement ("internal" strategy). A judge wanting to accomplish a certain external goal must attract enough colleagues to constitute a majority to affirm or reverse the lower court but also to agree on an opinion stating his views, a far more difficult task. He must be careful not to take so strong a position that he loses votes from the potential majority; conversely, a well-stated position may lead the justice writing the Court's position to adopt a colleague's language, and a persuasive dissent may even attract votes and lead to reversal of the Court's original position.[81]

Within the confines these constraints impose, the Court can engage in a variety of strategic actions. The first, docket management, involves deciding which cases to accept and which to reject. This is crucial so that the justices will not have to hand down the wrong decision at the wrong time and so that they will have the cases with the best factual setting for the rulings toward which they may be predisposed. In addition, by grouping several cases, as they did with such issues as school desegregation and the validity of confessions, they can focus attention on broad policy issues instead of the facts of a particular case. Docket management also includes making statements in the course of deciding cases that indicate the justices would welcome cases raising questions not presently before the Court and adopting substantive doctrinal rules to make clear to potential litigants that bringing certain types of cases will not be particularly fruitful. Rules on who has access to the courts or what cases are appropriate to be heard by judges are also of considerable strategic relevance. Certainly a ruling like that in *Baker v. Carr* that judges could rule on reapportionment complaints was an "open invitation" to press such suits.

The justices' disposition of cases they have accepted for review involves other strategic actions. One issue is whether to write a full, signed opinion or to use a summary action such as an affirmance of the lower court with only a citation of one or two cases. Such summary action is used to reinforce earlier decisions or to avoid full consideration of an issue where denial of review is thought inappropriate. Another question for the justices is the formal disposition to be given a case and the instructions, if any, to accompany the Court's mandate. These issues are closely related to the question of whether—and to what degree—to defer to lower court judges, particularly those in the state courts. Such deference, like that shown the legislative and executive branches, is necessary because the justices are dependent on lower court judges to carry out the Supreme Court's mandates and the lower court judges would be less likely to do so if the Supreme Court interfered frequently with their work. The deference may allow the lower courts to move ahead of the Supreme Court in applying a new doctrine, creating a base on which the Court itself can later build. However, it may also produce delay or noncompliance, with the case being brought back to the Court for further action and thus interfering with docket management. Particularly when the Court has remanded a case with only limited instructions, evasion may also result. This allows the lower court to find an alternative basis for reaching the result it had reached earlier while keeping the case out of the Supreme Court's hands.[82]

Other strategic options exist for the justices once they have decided to issue a full, signed opinion. They must take care to sound as much as possible like a "law court," drawing heavily on precedent and history where possible and avoiding appearing to base decisions on social science evidence, which tends to provoke criticism. Opinions vary in their clarity, in their breadth and narrowness, and in the directness with which issues are approached. Ambiguity may be used strategically, to help keep control of policy making in the Court's hands because unclear decisions force others to come back to the Court to obtain the justices' clarification. Thus ambiguous decisions, while producing "repeater" cases which may cause docket management problems, allow the Court to "monitor" the action of lower courts. With respect to the breadth of opinions, the Court can achieve its goals without broad opinions, particularly if negative reaction to a straightforward doctrinal ruling is foreseen; limited doctrinal grounds—basing a case on its facts or relying on statutory interpretation—can be used instead of dealing with larger, "tougher" constitutional questions. Yet there are times when reaching the principal issues directly is not only what the judges want to do but is also important strategically—to satisfy the demands and expectations of the Court's constituencies. If much of the Court's strategy seems based on avoiding issues or minimizing the degree to which they are confronted, one must remember that this is done so that on the issues the Court considers most important it can act forthrightly.

The Supreme Court is a major policy maker in the American political system. Its role as a policy maker is acknowledged by more and more people. Yet the strength of the myth that the Court finds and does not make the law and the special "legal" framework within which the justices operate mean that the degree to which its making of policy will be explicit and the degree to which strategy can be openly considered will in fact be limited. However, one must remember that the fact that the Supreme Court is a vital institution today is testimony to its ability in the long run to meet the expectations people have of it or its ability to act strategically—although that may mean moving to the center of the political arena at some times, while at other times finding it more appropriate to move closer to the periphery.

NOTES

1. For example, *Hurtado v. California*, 110 U.S. 516 (1884) (grand jury indictment), and *Twining v. New Jersey*, 211 U.S. 78 (1908) (Fifth Amendment).
2. *Abrams v. United States*, 250 U.S. 616 (1919), and *Schenck v. United States*, 249 U.S. 47 (1919). See also *Gitlow v. New York*, 268 U.S. 652 (1925).
3. On child labor legislation, see *Hammer v. Dagenhart*, 247 U.S. 251 (1918) (commerce), and *Bailey v. Drexel Furniture*, 259 U.S. 20 (1922) (taxation.)
4. For example, *Schechter Poultry Corp. v. United States*, 295 U.S. 495 (1935); *United States v. Butler*, 297 U.S. 1 (1936); *Carter v. Carter Coal Co.*, 298 U.S. 238 (1936).
5. *National Labor Relations Board v. Jones & Laughlin Steel Corp.*, 301 U.S. 1 (1937); *United States v. Darby*, 312 U.S. 100 (1941).
6. *Heart of Atlanta Motel v. United States*, 371 U.S. 241 (1964); *Katzenbach v. McClung*, 379 U.S. 294 (1964).
7. *National League of Cities v. Usery*, 96 S.Ct. 2465 (1976). Similarly, the dissenters in a recent Sherman Act case claimed that the Court was returning to substantive due process. *Cantor v. Detroit Edison Co.*, 96 S.Ct. 3110 at 3140 (1976).
8. *Smith v. Allwright*, 321 U.S. 649 (1944) (white primaries); *Shelley v. Kraemer*, 334 U.S. 1 (1948) (restrictive covenants); *Sweatt v. Painter*, 339 U.S. 629 (1950), and *McLaurin v. Board of Regents*, 339 U.S. 637 (1950) (law school and graduate education).
9. Glendon Schubert, *The Constitutional Polity* (Boston: Boston University Press, 1970), p. 71.
10. Reapportionment: *Reynolds v. Sims*, 377 U.S. 533 (1964); obscenity: *Roth v. United States/Alberts v. California*, 354 U.S. 476 (1957); libel: *New York Times v. Sullivan*, 376 U.S. 254 (1964); school prayer: *Engel v. Vitale*, 370 U.S. 421 (1962),

and *Abington School District v. Schempp,* 374 U.S. 203 (1963); public accommodations: *Heart of Atlanta Motel v. United States,* 371 U.S. 241 (1964), and *Katzenbach
v. McClung,* 379 U.S. 294 (1964); voting: *South Carolina v. Katzenbach,* 383 U.S.
301 (1966); housing: *Jones v. Mayer,* 392 U.S. 409 (1968). See generally Stephen
L. Wasby, *Continuity and Change: From the Warren Court to the Burger Court*
(Pacific Palisades, Calif.: Goodyear, 1976), pp. 71–77.

11. *Harper v. Virginia Board of Elections,* 383 U.S. 663 (1963); *Shapiro v. Thompson,*
     394 U.S. 618 (1960).

12. *Terry v. Ohio,* 392 U.S. 1 (1968); *Warden v. Hayden,* 387 U.S. 294 (1967); *Lewis
     v. United States,* 385 U.S. 206 (1966), and *Hoffa v. United States,* 385 U.S. 293
     (1966); *Berger v. New York,* 388 U.S. 41 (1967).

13. Lewis M. Steel, "Nine Men in Black Who Think White," *New York Times
     Magazine,* October 13, 1968, pp. 56, 117.

14. "The Role of the Supreme Court in a Democratic Society—Judicial Activism
     or Restraint?" *Cornell Law Quarterly* 54 (November 1968):8.

15. The California and New Jersey Supreme Courts provide judicial examples. The
     former invalidated the death penalty on the basis of California's constitution
     (although that ruling was overturned by the people of California when they
     amended their own constitution). It also held that a confession obtained without
     the *Miranda* warnings could not be used to impeach a witness's testimony,
     although the U.S. Supreme Court, interpreting the national Constitution, had
     allowed such use of such statements. In New Jersey, the high court invalidated
     the property tax as the method for financing public education, as had the
     California court earlier. See the comments by Justice Brennan, *American Bar
     Association Journal* 62 (August 1976):993–94. For an example of legislative action
     setting a higher criminal procedure standard than required by the U.S. Supreme
     Court, see Stephen Arons and Ethan Katsh, "Reclaiming the Fourth Amendment
     in Massachusetts," *Civil Liberties Review* 2 (Winter 1975):82–89.

16. Richard Funston, "Foreword: The Burger Court: New Directions in Judicial
     Policy-Making," *Emory Law Journal* 23(Summer 1974):656.

17. Robert G. McCloskey, "Reflections on the Warren Court," *Virginia Law Review*
     51 (November 1965):1234.

18. Paul Bender, "The Reluctant Court," *Civil Liberties Review* 2 (Fall 1975):101. For
     more recent terms, see John P. MacKenzie, "The Lost Court," *Civil Liberties
     Review,* 3 (October/November 1976): 36–53, and Stephen L. Wasby, "Certain
     Conservatism or Surprise: Civil Liberties in the 1976 Term," *Civil Liberties
     Review,* 4 (October/November 1977).

19. J. Woodford Howard, Jr., "Is the Burger Court a Nixon Court?" *Emory Law
     Journal* 23 (Summer 1974):762.

20. *Hudgens v. National Labor Relations Board,* 96 S.Ct. 1029 (1976), overruling *Food
     Employees v. Logan Valley Plaza,* 391 U.S. 308 (1968); *Greer v. Spock,* 96 S.Ct. 1211
     (1976), overruling *Flower v. United States,* 407 U.S. 197 (1972); *National League
     of Cities v. Usery,* 96 S.Ct. 2465 (1976), overruling *Maryland v. Wirtz,* 392 U.S.
     183 (1968); *Machinists v. Wisconsin Employee Relations Commission,* 96 S.Ct. 2548
     (1976), overruling *Automobile Workers v. Wisconsin Board,* 336 U.S. 245 (1949);

and *Andresen v. Maryland,* 96 S.Ct. 2737 (1976).

21. See the figures in James Simon, *In His Own Image: The Supreme Court in Richard Nixon's America* (New York: David McKay, 1973), pp. 110–31.
22. Glendon Schubert, "The Future of the Nixon Court," University Lecture, University of Hawaii, May 9, 1972, p. 12.
23. *Bishop v. Wood,* 96 S.Ct. 2074 (1976).
24. *Mahan v. Howell,* 410 U.S. 315 (1973); *Whitcomb v. Chavis,* 403 U.S. 124 (1974); but see *White v. Regester,* 412 U.S. 755 (1973).
25. *Swann v. Charlotte-Mecklenberg School District,* 402 U.S. 1 (1971) (busing); *Norwood v. Harrison,* 413 U.S. 455 (1973) (free textbooks to racially exclusive private schools); *Runyon v. McCrary,* 93 S.Ct. 2586 (1976) (admission to discriminatory private schools).
26. *Milliken v. Bradley,* 418 U.S. 717 (1974); *Hills v. Gautreaux,* 96 S.Ct. 1538 (1976).
27. *Moose Lodge v. Irvis,* 407 U.S. 163 (1973); *Palmer v. Thompson,* 403 U.S. 217 (1971); *Griggs v. Duke Power Co.,* 401 U.S. 424 (1971); *Albemarle Paper Co. v. Moody,* 422 U.S. 405 (1975); *Franks v. Bowman Transportation,* 96 S.Ct. 1251 (1976).
28. *Reed v. Reed,* 404 U.S. 71 (1971); *Frontiero v. Richardson,* 411 U.S. 676 (1973); *Taylor v. Louisiana,* 419 U.S. 522 (1975); *Cleveland Board of Education v. LaFleur,* 414 U.S. 632 (1974); *Roe v. Wade,* 410 U.S. 113 (1973), and *Doe v. Bolton,* 410 U.S. 179 (1973); *Planned Parenthood of Central Missouri v. Danforth,* 96 S.Ct. 2831 (1976).
29. *Geduldig v. Aiello,* 417 U.S. 484 (1974); *Kahn v. Shevin,* 416 U.S. 351 (1974).
30. *Goldberg v. Kelly,* 397 U.S. 254 (1970), and *Wheeler v. Montgomery,* 397 U.S. 280 (1970) (notice-and-hearing); *Dandridge v. Williams,* 397 U.S. 471 (1970) (maximum grants); *Wyman v. James,* 400 U.S. 309 (1971) ("home visit").
31. For higher education, see *Tilton v. Richardson,* 403 U.S. 673 (1971); *Hunt v. McNair,* 413 U.S. 634 (1974); and *Roemer v. Board of Public Works,* 96 S.Ct. 2337 (1976). Representative decisions at the elementary and secondary education level include *Lemon v. Kurtzman,* 403 U.S. 602 (1971) (*"Lemon I"*); *Levitt v. Committee for Public Education & Religious Liberty,* 413 U.S. 472 (1973); and *Meek v. Pittenger,* 421 U.S. 349 (1975).
32. *Walz v. Tax Commission,* 397 U.S. 664 (1970); *Wisconsin v. Yoder,* 406 U.S. 205 (1972).
33. *New York Times v. United States,* 403 U.S. 713 (1971); *Nebraska Press Association v. Stuart,* 96 S.Ct. 2791 (1976); *Branzburg v. Hayes,* 408 U.S. 665 (1972); *Parker v. Levy,* 417 U.S. 733 (1974); and *Greer v. Spock,* 96 S.Ct. 1211 (1971).
34. *United States v. Calandra,* 414 U.S. 338 (1974); *United States v. Janis,* 96 S.Ct. 3021 (1976); *Stone v. Powell,* 96 S.Ct. 3037 (1976).
35. *Almeida-Sanchez v. United States,* 413 U.S. 266 (1973); *United States v. Robinson,* 414 U.S. 218 (1973); *Cady v. Dombrowski,* 413 U.S. 433 (1973); and *Schneckloth v. Bustamonte,* 412 U.S. 218 (1973); *California Bankers Association v. Schultz,* 416 U.S. 21 (1974), and *United States v. Miller,* 96 S.Ct. 1619 (1976); *Couch v. United States,* 409 U.S. 322 (1973), and *Fisher v. United States,* 96 S.Ct. 1569 (1976); *Andresen v. Maryland,* 96 S.Ct. 2737 (1976).
36. *Kastigar v. United States,* 406 U.S. 441 (1972).

37. *Michigan v. Mosley*, 96 S.Ct. 321 (1975); *Harris v. New York*, 401 U.S. 222 (1971), and *Oregon v. Hass*, 420 U.S. 714 (1975); *United States v. Mandujano*, 96 S.Ct. 1612 (1976); *Beckwith v. United States*, 96 S.Ct. 1612 (1976).

38. *Argersinger v. Hamlin*, 407 U.S. 25 (1972); *Kirby v. Illinois*, 406 U.S. 682 (1972) (line-up); *Ross v. Moffitt*, 417 U.S. 600 (1974) (discretionary appeals); *Morrissey v. Brewer*, 408 U.S. 471 (1972) (parole); *Gagnon v. Scarpelli*, 411 U.S. 778 (1973) (probation); *Wolff v. McDonnell*, 418 U.S. 539 (1974) (prison discipline); *Fuller v. Oregon*, 417 U.S. 40 (1974) (recoupment statutes).

39. *North Carolina v. Alford*, 400 U.S. 25 (1970), and *Santobello v. New York*, 404 U.S. 257 (1971); *Brady v. United States*, 397 U.S. 742 (1970).

40. Max Lerner, "Constitution and Court as Symbols," *Yale Law Journal* 46 (1939): 1294–95.

41. Martin Shapiro, *Law and Politics in the Supreme Court* (New York: Free Press, 1964), p. 328.

42. 384 U.S. 436 at 531–32 (1966); emphasis supplied.

43. *West Virginia State Board of Education v. Barnette*, 319 U.S. 624 at 647 (1943).

44. *Furman v. Georgia*, 408 U.S. 238 at 405–406, 410–11 (1972).

45. *U.S. v. Midwest Video Corp.*, 406 U.S. 649 at 676 (1972).

46. Robert Dahl, *Democracy in the United States* (New York: Random House, 1972), p. 201.

47. Alexander Bickel, *The Least Dangerous Branch* (Indianapolis: Bobbs-Merrill, 1962), p. 58.

48. Herbert Wechsler, "Toward Neutral Principles of Constitutional Law," *Harvard Law Review* 73 (November 1959): 19, 11.

49. Bickel, *The Least Dangerous Branch*, p. 59. See also Bickel, "The Supreme Court's 1960 Term, Foreword: The Passive Virtues," *Harvard Law Review* 75 (November 1961): 40–79.

50. See *Sheppard v. Maxwell*, 384 U.S. 333 (1966), and *Nebraska Press Association v. Stuart*, 96 S.Ct. 2791 (1976).

51. See the comment of Justice Stewart, *Gregg v. Georgia*, 96 S.Ct. 2909 at 2926 (1976).

52. The three cases are *Food Employees v. Logan Valley Plaza*, 395 U.S. 575 (1968); *Lloyd Corp. v. Tanner*, 407 U.S. 551 (1972); and *Hudgens v. National Labor Relations Board*, 96 S.Ct. 1029 (1976).

53. *National League of Cities v. Usery*, 96 S.Ct. 2465 (1976), overruling *Maryland v. Wirtz*, 392 U.S. 183 (1968).

54. Compare *Terry v. Ohio*, 392 U.S. 1 (1968), with *United States v. Robinson*, 414 U.S. 218 (1973), and *Gustafson v. Florida*, 414 U.S. 260 (1973).

55. *United States v. Wade*, 388 U.S. 218 (1967); *Gilbert v. California*, 388 U.S. 263 (1967); *Kirby v. Illinois*, 406 U.S. 682 (1972); *United States v. Ash*, 413 U.S. 300 (1973).

56. See *Francis v. Henderson*, 96 S.Ct. 1708 at 1712 (1976) (Justice Brennan), and *Ludwig v. Massachusetts*, 96 S.Ct. 2781 at 2789 (1976), where Justice Stevens criticized the Court for "refusing to follow a precedent so nearly in point."

57. *Runyon v. McCrary*, 92 S.Ct. 2586 at 2601–602 (Justice Powell), 2603–604 (Justice Stevens).

58. Walter F. Murphy, *Elements of Judicial Strategy* (Chicago: University of Chicago Press, 1964), p. 204.

59. Martin Shapiro, "Stability and Change in Judicial Decision-Making: Incrementalism or Stare Decisis?" *Law in Transition Quarterly* 2 (Summer 1965): 155.

60. Fred Graham, *The Self-Inflicted Wound* (New York: Macmillan, 1970), p. 171.

61. See *Manual Enterprises v. Day*, 370 U.S. 478 (1962); *Jacobellis v. Ohio*, 378 U.S. 194 (1964); and the trilogy of *Mishkin v. New York*, 383 U.S. 502 (1966), *Ginzburg v. United States*, 383 U.S. 463 (1966), and *Memoirs v. Massachusetts*, 383 U.S. 413 (1966).

62. *Baker v. Carr*, 369 U.S. 186 at 267–70 (1962).

63. Concurring, *Ashwander v. Tennessee Valley Authority*, 297 U.S. 288 at 346–48 (1936).

64. Simon, *In His Own Image*, pp. 8, 227.

65. *United States v. Carolene Products*, 304 U.S. 144 (1938).

66. See Justice Stewart's comments in the most recent death penalty cases, *Gregg v. Georgia*, 96 S.Ct. 2909 at 2925–26 (1976).

67. *Rice v. Sioux City Memorial Cemetery*, 349 U.S. 71 (1955); *Jackson v. Alabama*, 348 U.S. 888 (1954). Full discussion of this point and most of the argument in this section can be found in Stephen L. Wasby, Anthony A. D'Amato, and Rosemary Metrailer, *Desegregation from Brown to Alexander: An Exploration of Supreme Court Strategies* (Carbondale, Ill.: Southern Illinois University Press, 1977).

68. Anthony D'Amato and Robert O'Neil, *The Judiciary and Vietnam* (New York: St. Martin's Press, 1972).

69. S. Sidney Ulmer, "Parabolic Support for Civil Liberties," *Florida State University Law Review* 1 (Winter 1973): 149.

70. Murphy, *Elements of Judicial Strategy*, p. 193.

71. S. Sidney Ulmer, "Earl Warren and the Brown Decision," *Journal of Politics* 33 (August 1971): 697.

72. Alpheus Thomas Mason, *Harlan Fiske Stone: Pillar of the Law* (New York: Viking Press, 1956), pp. 614–15.

73. Chief Justice Burger, 409 U.S. xxxv at xvi (1972). The case was *Taylor v. Georgia*, 315 U.S. 25 (1940).

74. *San Antonio School District v. Rodriguez*, 411 U.S. 1 at 56–58 (1973).

75. *Elkins v. United States*, 364 U.S. 206 (1960); *Gregg v. Georgia*, 96 S.Ct. 2909 at 2628 (1976), and *Woodson v. North Carolina*, 96 S.Ct. 2978 at 2988–89 (1976); *Linkletter v. Walker*, 381 U.S. 618 (1965).

76. G. Gregory Fahlund, "Retroactivity and the Warren Court: The Strategy of a Revolution," *Journal of Politics* 35 (August 1973): 570–93.

77. Rohde found that majorities were larger in the face of threats than at other times. David Rohde, "Policy Goals and Opinion Coalitions in the Supreme Court," *Midwest Journal of Political Science* 16 (May 1972): 218–19.

78. Murphy, *Elements of Judicial Strategy*, pp. 4–5.

79. Ibid., p. 4.

80. Ibid., p. 158.

81. See Alexander Bickel, ed., *The Unpublished Opinions of Mr. Justice Brandeis* (Cambridge, Mass.: Harvard University Press, 1957), for examples. Another instance

occurred when Justice Byrnes's draft dissent in a Fair Labor Standards Act case persuaded Justice Jackson to change his vote, with the Byrnes draft being issued as the Court's majority opinion. *Walling v. Belo Corp.*, 316 U.S. 624 (1941). See 409 U.S. xxxv at liv.

82. See Note, "State Court Evasion of United States Supreme Court Mandates," *Yale Law Journal* 36 (1947): 574–83, and Jerry K. Beatty, "State Court Evasion of United States Supreme Court Mandates During the Last Decade of the Warren Court," *Valparaiso University Law Review* 6 (Spring 1972): 260–85.

2 | JUDICIAL REVIEW AND PUBLIC OPINION

The courts of all nations have to interpret legislation and its implementation by the executive. However, in relatively few can they exercise the American version of judicial review—or "constitutional review"[1]—the determination of whether a statute or other action is consistent with the Constitution. While in some nations separate courts are established to deal with constitutional questions, in the United States such questions are posed not as abstract issues by those seeking advice apart from a specific factual situation but are answered in the course of regular litigation. Furthermore, all American courts, not merely the high appellate courts, have the power to engage in judicial review, which they exercise concurrently with their other tasks. Although we do not use separate "constitutional courts," the proportion of the Supreme Court's caseload taken up with constitutional questions has led some observers to say that the Court has become much like one.

The exercise by the courts of the power of judicial review is important not only because of the constitutional doctrine the courts thus establish. Where the courts stand in public opinion is also substantially affected by their exercise of judicial review. This is particularly true with respect to the Supreme Court for at least two reasons. One is that while lower courts may invalidate statutes, the Supreme Court is usually called upon to sustain or overturn such decisions; people look to it for a final judicial say on the matter. The other is that the public tends to remember the Supreme Court's actions in the course of judicial review—particularly those limiting or "checking" the other branches—more than it does the Court's other actions such as those in which it interprets statutes. The Court's standing with the

general public and particularly its more attentive publics, thus affected by
the Court's own actions, in turn at least in part affects its further exercise
of judicial review and the way it carries out its role. Public opinion is an
important part of the Court's environment affecting the judges' strategies
discussed in the previous chapter, by and large serving as a limit on what
the Court can do.

JUDICIAL REVIEW

Establishment and Early Development

The development of judicial review in the United States was affected by
political philosophy, such as Montesquieu's theory of the separation of
powers. Also relevant, although not determinative, was the practice concerning the review of decisions made in the American colonies. "Conformity
clauses" in state charters required acts of the colonial legislatures to conform
to the laws of England. They gave the king's Privy Council and its Committee on Appeals the power to disallow the legislation—power exercised in
fact by the Board of Trade. Such action was, however, review without a
written constitution, which England did not have, so that it did not lead
those in the colonies to associate judicial review with a written constitution.
Decisions of the courts in the colonies could also be appealed to the Privy
Council when the royal governor granted a request to appeal. Here the
Committee on Appeals often based its decisions on the colony's local law.

The Articles of Confederation did not establish judicial review, as the
Articles contained no provision for a separate national judiciary. Congress
became the court of last appeal, or the board of arbitration, for controversies
between the states and for a limited number of cases involving individual
rights. The Constitution's Supremacy Clause was anticipated, however, in
the stipulation that the Articles and acts of Congress be accorded the status
of law within the states:

> Every State shall abide by the determination of the United States in Congress
> assembled, on all questions which by this Confederation are submitted to them.
> And the Articles of this Confederation shall be inviolably observed by every
> State . . . (Art. XIII)

No plan submitted to the Constitutional Convention of 1787 conferred
directly upon the judiciary any power of passing on the constitutionality
of congressional acts. Under one major proposal—the Virginia Plan—there
would have been a national court system whose judges would have been
chosen by Congress. Congress would have had the right to disallow state

---

FROM THE UNITED STATES CONSTITUTION
The judicial power of the United States shall be vested in one Supreme Court,
and in such inferior Courts as the Congress may from time to time ordain and
establish. (Art. III, Sec. 1)

The judicial power shall extend to all cases, in law and equity, arising under
this Constitution, the laws of the United States, and treaties made, or which shall
be made, under their authority; to all cases affecting ambassadors, other public
ministers and consuls; to all cases of admiralty and maritime jurisdiction; to
controversies to which the United States shall be party; to controversies between two or more states; between a state and citizens of another state; between
citizens of different states; between citizens of the same state claiming lands
under grants of different states; and between a state, or the citizens thereof, and
foreign states, citizens or subjects.
    In all cases affecting ambassadors, other public ministers and consuls, and
those in which a state shall be party, the supreme court shall have original
jurisdiction. In all the other cases before mentioned, the supreme court shall
have appellate jurisdiction, both as to law and fact, with such exceptions and
under such regulations as the Congress shall make. (Art. III, Sec. 2)

This Constitution, and the Laws of the United States which shall be made
in Pursuance thereof and all Treaties made, or which shall be made, under the
Authority of the United States, shall be the supreme Law of the Land; and the
Judges in every State shall be bound thereby, any Thing in the Constitution
or Laws of any State to the Contrary notwithstanding (Art. VI, cl.2)

---

legislation; a council of revision composed of the executive and a "conven-
ient number" of the judiciary would have had a suspensive veto over national legislation. The major alternate plan made acts of Congress and treaties the "supreme law of the respective States." The state judiciary would
be bound by congressional acts, notwithstanding state constitutional provisions. Because state courts would initially decide federal cases, the need for
lower federal courts would be eliminated and the strength of the national
government thus decreased. There would, however, be a supreme national
tribunal, appointed by the executive. This court would have appellate jurisdiction in certain classes of cases coming from the state courts.
    The Convention eliminated congressional disallowance of state legislation
and judicial participation in the veto. Establishment of a Supreme Court was
indicated, with Congress having the power to create lower federal courts.
Although the jurisdiction of the courts generally and of the Supreme Court
in particular was spelled out, and although supremacy of national law was
established, not a word was said regarding the power of these federal courts

to invalidate laws contrary to the Constitution. As with much of the rest of the document, the generality of the language, which left matters to later development, was recognized but was felt necessary to obtain approval of the Constitution.

The leaders of the Constitutional Convention apparently did not foresee that judges would engage in general expounding of the Constitution but did believe that the federal judiciary had the right to refuse to recognize unconstitutional federal law. Hamilton, Madison, and Jay, who wrote *The Federalist* papers to secure ratification of the Constitution in New York, argued strongly for judicial review. (They dealt summarily with the Supreme Court's right to overrule state legislation, thus indicating it was less of an issue, at least for them.) According to *The Federalist*, the Constitution as fundamental law was to be preferred when there was an irreconcilable variance between it and a legislative act, and the courts had a duty "to declare all acts contrary to the manifest tenor of the Constitution" void. There was, however, no direct empowerment to construe laws according to the Constitution's *spirit*. Judicial review was not an exercise of judicial power over the legislative branch. Instead, through judicial review the intention of the people—who ratified the Constitution—would be enforced against the intention of their agents, the legislators; the prior act of a superior authority would be preferred to the subsequent act of an inferior one.

The authors of *The Federalist* also noted that state judges were to be incorporated into the operation of the national government and would be bound by oath to support federal laws when those laws concerned legitimate and enumerated objects of federal jurisdiction. The Supremacy Clause was deemed imperative in helping produce needed uniform interpretation of federal law and treaties, as was national court jurisdiction and a supreme tribunal of last resort, because without appeal and review, state courts of final jurisdiction would produce an endless variety of decisions on the same point. So that all matters of national law would receive original or final determination in the national courts, appeal would run from state courts to either the federal district courts or the Supreme Court. However, in an appeal to states' rights interests, it was pointed out that the state courts—which would look beyond their own law in making decisions and would recognize relevant federal court rulings—would retain their former jurisdiction unless specifically limited by Congress, and that this system of initial state court determination with appeals from the state court to the federal courts would actually diminish the number of federal courts.

Judicial review of state decisions occurred before judicial review of national actions and, despite controversy, became firmly established earlier. In 1797, in *Ware v. Hylton,* the Court held our treaty of peace with Great Britain superior to Virginia's statute sequestering property. Thirteen years later, in *Fletcher v. Peck,* stemming from the Yazoo land fraud case, the Court ruled a Georgia statute in violation of the Constitution.[2] The most severe early test of judicial review of state decisions came with respect to the Supreme Court's power over actions of state courts when the Virginia courts declared unconstitutional the federal Judiciary Act provision establishing appeal of state decisions affecting federal rights. Responding, Justice Joseph Story ruled in *Martin v. Hunter's Lessee* (1816) that the U.S. Supreme Court had a right to review the decisions of state courts in order to produce uniform interpretation of the nation's "supreme law." Five years later, Virginia's claim that an appeal by a criminal defendant from a state ruling was a violation of state sovereignty was rejected by Chief Justice Marshall in *Cohens v. Virginia*. Marshall stated that the Supreme Court had the right to appellate jurisdiction over decisions of highest state courts in all questions of national power.

Marbury v. Madison, the first instance in which the Court invalidated a national statute, did not come until 1803. Although the Court had earlier reviewed and upheld acts in Congress,[3] it was the *Marbury* decision which established judicial review. In the waning hours of the Federalist administration, Marbury had been appointed to a minor judicial position but had not received his commission. He sought a writ of mandamus from the Supreme Court to make the new secretary of state, James Madison, hand over the commission. Chief Justice John Marshall, himself one of the last of the Federalist appointees, found that Marbury had a right to his commission and thus was entitled to a remedy. However, he ruled that the Court could not issue the requested writ because the Judiciary Act of 1789 unconstitutionally added to the Court's original jurisdiction:

> If congress remains at liberty to give this court appellate jurisdiction, where the Constitution has declared their jurisdiction shall be original; and original jurisdiction where the constitution has declared it shall be appellate; the distribution of jurisdiction, made in the constitution, is form without substance. . . .

Marshall thus gave himself an opportunity to lecture the Jeffersonians and established judicial review over acts of Congress, perhaps putting the administration on notice that future legislation might be declared unconstitutional.

Marshall's opinion is open to considerable criticism. Under today's standards, he would have withdrawn from the case because of his earlier involvement; as secretary of state, he had failed to deliver Marbury's commission. Had he followed usual procedure and looked first at whether the Court had jurisdiction, he could have invalidated the statute but could not have lectured the administration about not giving up the commission. More important, Marshall could have followed a rule of judicial self-restraint and construed the Act of 1789 to save its constitutionality by saying the items of original jurisdiction enumerated in the Constitution need not be exclusive but could be supplemented by Congress. Such criticisms, however, ignore the long-term implications of the case. The invalidation of the statute and what Marshall said about judicial review give the case its importance. Marshall insisted that the language of the Constitution be taken seriously. "It is a proposition too plain to be contested," he asserted, "that the constitution controls any legislative act repugnant to it; or that the legislature may alter the constitution by an ordinary act"; there were no other alternatives. The Constitution had to be either a "superior paramount law, unchangeable by ordinary means" or it was like any other law. But, Marshall said,

> Certainly, all those who have framed written constitutions contemplate them as forming the fundamental and paramount law of the nation, and consequently, the theory of every such government must be, that an act of the legislature repugnant to the constitution is void.

From this he concluded that the courts should invalidate the "repugnant" acts, giving the courts power over the other branches of government when constitutional questions were at issue.

This conclusion was not, however, the only one he could have reached. Judge Gibson of the Pennsylvania Supreme Court was later to say in *Eakin v. Raub* that the courts were limited to looking to see whether the legislature, in passing a law, had used the proper procedures because "the legislative organ is superior to every other, inasmuch as the power to will and command is essentially superior to the power to act and to obey." Judges implementing a law properly passed but in violation of the Constitution were not, Gibson said, themselves committing an unconstitutional act but only doing what they were required to do.

One could also argue that each branch of government had to decide for itself what was constitutional and had to act on the basis of its own conclusions. For example, in vetoing legislation for a national bank already held to be constitutional, President Andrew Jackson said that he had to make his own judgment on the matter. And President Jefferson had thought that the Supreme Court could make determinations as to the validity of acts of Congress but that these determinations, while entitled to respect, were not binding on the president. The position that each branch had to decide for itself also meant that the courts could not be forced to enforce policies thought by the judges to be unconstitutional. Giving the courts somewhat more authority was a position that they should have the power of judicial review over matters within courts' special competence or directly affecting the judiciary, such as the jurisdictional question in *Marbury*.

Marshall, of course, went further. In language the Supreme Court used recently in ruling in *United States v. Nixon* that the president was not the sole judge of when he could invoke executive privilege, Marshall said:

It is, emphatically, the province and duty of the judicial department, to say what the law is. Those who apply the rule to particular cases must of necessity expound and interpret that rule. . . . If a law be in opposition to the constitution; if both the law and the constitution apply to a particular case, so that the court must either decide that case, conformably to the law, disregarding the constitution, or conformably to the constitution, disregarding the law; the court must determine which of these conflicting rules governs the case; this is of the very essence of judicial duty.

Marbury has come to stand for the Supreme Court's power—and, by extension, the power of other courts—to invalidate acts of Congress and the president. Yet, although the doctrine was extended the following year to executive orders,[4] the power was not used again until 1857, when the Court in *Dred Scott v. Sandford* struck down congressional action concerning slavery in the territories. Its use became common only at the end of the nineteenth century and it was used frequently only in the 1930s. Yet, despite the relative infrequency of the Court's invalidation of congressional and executive acts, judicial review has always been available to the judges. This should remind us that, in Justice Benjamin Cardozo's words, "The utility of an external power restraining the legislative judgment is not to be measured by counting the occasions of its exercise."[5]

Arguments about Judicial Review

All controversy about Supreme Court decisions is in some measure result-oriented, that is, a reaction to the substance of rules being sustained or invalidated. However, when the Court strikes down state laws, people are particularly likely to argue on the basis of their feelings about the statutes themselves, whether on school segregation or search and seizure procedures, rather than to express a commitment to a principle that the Supreme Court should not review any state decisions. Indeed, the need for uniform national interpretation of constitutional provisions and of national law has been generally accepted. It has even led to the feeling that the Supreme Court's power to review state decisions is more important than its power over national government actions. As Justice Holmes put it, "I do not think the United States would come to an end if we lost our power to declare an act of Congress void. I do think the Union would be imperilled if we could not make that declaration as to the laws of the several states."[6] And throughout our history the Supreme Court has invalidated state laws more frequently than federal laws. Until the end of the Civil War, while only two federal laws were invalidated by the Court, sixty state laws were struck down. In the period from 1888 to 1937, over seventy national statutes were declared invalid while over four hundred state laws were struck down. In the most recent period, from 1953 to 1976, twenty-five federal statutes were voided while over a thousand state laws were similarly eliminated.[7]

Situations of federal-state conflict clearly show the need to have the authority to invalidate state law in aid of uniform national law. In the period prior to the Civil War, in connection with the fugitive slave issue, the Court had to rule in *Ableman v. Booth* (1842) that Congress's jurisdiction over fugitive slaves was exclusive and that the existence of a federal statute on the subject required invalidation of a conflicting state law. In *Prigg v. Pennsylvania*, the Court also had to deny a state court's right to order federal officials to commit an act in conflict with a federal court decision. During the height of resistance to the Supreme Court's desegregation decisions, the governor of Arkansas asserted that a Supreme Court decision was not part of the "supreme law of the land," so that state governments were not bound by it. In the Little Rock case, *Cooper v. Aaron,* the Supreme Court rejected this contention, with the justices going out of their way to stress the supremacy of national law, including the Supreme Court's interpretation of it.

Controversy has been quite likely when the Court has ruled that federal activity prevents the states from taking action on a particular subject because the national government has preempted the field, as in the fugitive slave matter just noted.[8] More recently, in *Pennsylvania v. Nelson* (1956), the Supreme Court set aside a state conviction of a person for attempting to overthrow the government of the United States because the national government had indicated its intention—through enactment of the Smith and McCarran Acts—to preempt the field. Intense negative reaction to this decision—and to several others in the internal security area—led the Court to rule in *Uphaus v. Wyman* (1959) that the states could regulate subversion directed at the states themselves. Preemption through the treaty-making power has also stirred the anger of those advocating "states' rights." One such case was *Missouri v. Holland* (1920), in which the Court ruled that a federal treaty (with Canada) concerning migratory waterfowl supplanted Missouri's hunting regulations. The Court's statement that in implementing a treaty the national government could go beyond the constitutional powers it would have in the absence of a treaty led to later attempts to amend the Constitution to limit the president's treaty-making power.

Despite such controversy or the unhappiness provoked when the Court "intrudes" into particular policy areas where the states have strong commitments, most debate about judicial review, which has focused on its application to the other branches of the national government, has centered on its general wisdom. Such debate, which *Marbury* began, has been based more on principle than has argument over invalidation of state actions. The controversy reaches far more broadly than quarrels about the wisdom of the Court's use of judicial review in particular cases or whether the Court's solution to a problem produced the consequences the Court intended. One type of argument revolves around whether courts are more appropriate than other governmental institutions for dealing with certain problems.[9] In making this judgment, one could use several criteria: whether judges are familiar with conditions causing the problem, the language in which the problem is stated, and consequences of possible solutions; whether courts after initial action can reformulate policy based on information from implementation efforts; and whether the public believes in the courts' authority and competence to handle the problem—and whether the courts are more capable than other institutions in these respects.

Another, more frequently heard argument concerns whether judicial review is democratic. [10] On the one hand, Supreme Court judicial review is said to be undemocratic because justices appointed for life, who often remain on the Court long after they are no longer in tune with the country's views, can overrule acts of representatives elected periodically and of the executive branch whose head, the president, is also elected. There are several responses made to this argument. One is that the Court's statutory rulings can be overturned by congressional action alone and that its constitutional rulings can be reversed by constitutional amendment. This has occurred with the Eleventh Amendment (no suits against the states by citizens of another state), the Sixteenth Amendment (income tax), the Twenty-sixth Amendment (18-year-old vote), and the post-Civil War amendments eliminating slavery and redefining citizenship.[11] Another response, Hamilton's argument, is that through judicial review the Constitution—the people's will —is enforced over the will of the representatives. A third is that courts must protect minority rights, as much a part of democracy as is majority rule. Judicial review is also said to inject important "sober second thoughts" into the political decision-making process.

The argument that judicial review is undemocratic may be misdirected. For one thing, other governmental institutions may be no more responsive to the public than is the Court, with no one (except perhaps the president) accountable to a majority of the nation's voters. Moreover, our government's institutions operate by means of a system of checks and balances in which each participates in the work of the others and thus limits them. The Court's involvement in these political processes, engaging in strategic actions as a policy maker, thus makes it to some extent a democratic institution; only if it were fully insulated from those processes and if it found the law in total independence of other government bodies would it be undemocratic.

Such arguments do not assuage fears that judicial review is undemocratic, and those fears appear in debate about whether the Court legitimizes the work of the other elected branches of government or interferes with their actions. In what has for some time been the accepted position, Robert Dahl asserted that the Court does not block the legislative and executive branches. Instead, the Court is a part of the dominant national alliance and seldom is outside that alliance for long. Without support from president and Congress, the Court generally can do little. The Court might win small battles, but the justices are not likely to be successful "on matters of *major* policy, particularly if successive presidents and Congresses continue to support the policy the Court has called unconstitutional." When other political actors are unable to decide important questions, the Court can take action, but even then it does so at great risk and can succeed only if its actions are in tune with the political leadership's norms. Thus, "ordinarily the main contribution of the Court is to confer legitimacy on the fundamental policies of the successful coalition" governing the country.[12]

In making his argument, Dahl relied on the seventy-eight cases through 1957 in which the Supreme Court had exercised judicial review to invalidate eighty-six provisions of federal law. He also considered the timing of that review. Only half of the Court's actions came within four years of the legislation's enactment, with more than one-third of the "prompt" overrulings occurring during the Court's emasculation of the New Deal. In most instances when recent legislation was overturned, Congress reversed the Court or the Court itself did so somewhat later.

Dahl's position has been met directly by David Adamany and Jonathan Casper. Adamany's argument is that Dahl overestimated the Court's ability to legitimize. Casper's view, discussed below, is that Dahl underestimated the Supreme Court's effectiveness. The Court cannot legitimize, Adamany says, unless it strikes down the acts of other branches at least occasionally. If all statutes and administrative actions were sustained, the Court's ability to legitimize would be meaningless. During crises, the Court cannot grant legitimacy because it is at such times that justices are of the party opposite to that controlling the elected branches, and the Court's actions invalidating legislation certainly do not grant legitimacy to the new coalition. (Judicial review has been more likely when one political party has dominated Congress and the other party controlled the Court, with judicial review "significantly shaped by such political considerations as the degree of party difference between Congress and the Court, the nature of the party in power in the national government, and the party affiliations of the individual judges deciding specific cases."[13] Reexamination of judicial review during periods of partisan realignment has, however, led to the finding that, with the New Deal period eliminated because of its extremely high rate of Supreme Court invalidation of recent legislation, the Court was actually more likely to invalidate a law in periods of partisan stability than during partisan shift.[14]) Adamany says that conflict between the Court and the elected branches after realigning elections—those in which party control of the presidency changes and there is substantial shift in the voters' partisan idenfication—has "somewhat discredited and sometimes checked the lawmaking majority." In each such situation, however, the result has been "a clash that left doubtful the Court's capacity ultimately to legitimize the new regime and its policies." One hardly has legitimization when as a result of changes in personnel or its own strategic movement the Court finally adopts the dominant coalition's position. Instead there is "more the appearance of surrender to superior force."[15]

Adamany also points out that the public does not seem to know about the Court's legitimizing role; what little the public does know is principally about cases in which majorities have been checked, not legitimized. The Court's decisions also do not eliminate substantial opposition to the policies it has enunciated. Moreover, the views of political elites, who know more than does the general public about the Court and its decisions, are largely conditioned by the elites' political attitudes, not the Court's actions. Adamany concludes that the Court does not legitimize policy for the nation through the elites; elites confer legitimacy on the courts. The Court does not itself help create respect for the political system but commands respect, he says, because it is part of the political system. This ultimately saves the Court when a new political coalition considers limiting the justices. "Elements of the coalition's elected elite and of its electoral base now hold back in reverence to a constitutional institution, whose actions and function they may or may not fully understand or approve."[16] Here it is important to note that while "decisions of the Court . . . provide strong ammunition for congressmen who would defend the status quo—as defined by the Court, of course—against proposals for legislative change,"[17] they do not prevent legislative attack on the Court, including efforts to reverse what the Court has done. However, it has been shown that "reversal bills" were less likely to pass when arguments about the Court's sacrosanct nature were used more frequently and broad "anti-Court" attacks do produce more use of the argument that the Court should remain inviolate than do "decision-reversal" proposals.[18]

On the basis not only of the period used by Dahl but also on the basis of the Court's behavior since 1957, Casper argues that the Court has intervened decisively in the policy-making process. He points out that from 1957 through 1974 the Court held thirty-two provisions of federal law unconstitutional in twenty-eight cases. This means that more than one-fourth of the 106 cases in which national legislation has been invalidated have occurred within the last twenty years. Legislation was declared invalid within four years of enactment in only six of the recent cases and in only one of those, on the eighteen-year-old vote, was the Court reversed (by constitutional amendment); in none of the cases involving older legislation was the Court's action overturned. Casper seems to agree with Dahl that the Court's intervention in policy making is most likely to be effective when political elites

—and the general public—are substantially divided. Here, too, while contro-versy has followed the Court's rulings, efforts to overturn the Court's action have also failed.

Casper also asserts that the Supreme Court is seen as even more effective if one recognizes that even when its position is ultimately overturned, the Court may delay the effectiveness of policies for some years after they are enacted, and if one takes into account judicial actions Dahl ignored. Casper would include cases of statutory construction through which the justices have produced important policy on such subjects as welfare residence re-quirements and "man-in-the-house" rules, excusing conscientious objectors from military service, broadcasters' rights to refuse editorial commercials, and surveillance without a warrant. The Supreme Court's initiatives not involving invalidation of national legislation, such as Marshall's *McCulloch v. Maryland* and *Gibbons v. Ogden* rulings, must also be taken into account, as should the Court's introduction of new issues and new participants into the political process, which may assist actors' ability "to attract adherents, mobi-lize resources, and build institutions."[19]

THE SUPREME COURT AND PUBLIC OPINION

Early History and Editorial Reaction

In earlier times, even though we did not have available public opinion polls and other relatively sophisticated instruments for recording public opinion, it was clear that the Supreme Court's decisions had an impact on opinion. The decisions often produced mixed reactions. When the Court invalidated state fugitive slave statutes, "the decision was equally unsatisfac-tory to both pro-slavery and anti-slavery men" because the former were upset at the blow to states' rights and the latter, who disliked the federal fugitive slave law, thought the Court was backing the South.[20] The Court also suffered in the public eye when, shortly after being enlarged and only a year after it had held in the first Legal Tender Case that the Union government could not require that debts made before the passage of the Legal Tender Act be paid in paper money, the Court reversed itself. Al-though the legal community felt that the Legal Tender Act was constitu-tional, the reopening of the case was "a mistake which for many years impaired the people's confidence, not in the honesty, but in the impartiality and good sense of the Court."[21]

In the New Deal period, divided reaction was again evident. When the Court invalidated the National Industrial Recovery Act, "the more conser-vative sections of the press welcomed it as putting an end to unsound experiments in government regulation of industry," labor opposed the decision, and the business community was divided.[22] However, more and more elements of the public were alienated by the Court's continued striking down of New Deal legislation, and "each new adverse decision in the winter and spring of 1936 brought new bursts of hostility."[23] The farmers were upset about invalidation of the Agricultural Adjustment Act, the voiding of the Bituminous Coal Act irritated workers, and the minimum wage rulings "alienated nearly everybody"—including supporters of earlier decisions. Only ten of 344 editorials approved the decision on the minimum wage, with some sixty papers, including a number of conservative ones, calling for a constitutional amendment on the subject.

Editorial reaction to decisions is one measure, although an indirect one, of public opinion, and it has thus received some attention. Twenty-four large-circulation newspapers generally favored separation of church and state after three of four major Supreme Court church-state decisions; the exception was the *Zorach* ruling upholding New York's released time program, which the papers favored. The *McCollum* ruling striking down religious classes on school property was the most favorably received (eight papers favoring and two opposing), with editorials on the school prayer case closely divided (thirteen favoring, nine opposing). Support for the released time ruling may have resulted from that program's milder link between church and state, more conservative public attitudes toward civil liberties in 1952, liberal Justice Douglas's authorship of the Court's opinion, and the general tendency of newspapers to support the Court. Newspapers' editorial positions on church-state matters were affected by the city's political climate (the more Democrats, the more likely a paper to favor church-state separation), the publisher's politics (same relationship, but stronger), and the publisher's religion (higher support for the Court if Catholic or Episcopalian).[24]

A study of a much larger number of papers—sixty-three—revealed that twenty-seven had opposed the school prayer ruling in editorials while sixteen had favored it, and editorial cartoons were also more likely to be critical of the ruling than to favor it. These proportions were more in line with general public opinion than the more favorable reaction of the big-city papers just noted. Opposition to the ruling was strongest in the upper Midwest, with more southern papers neutral or favorable. Editorial reaction to the Court's first reapportionment decision was generally favorable, with thirty-eight editorials favoring, ten opposing, and another twelve neutral or confused; sixteen editorial cartoons were favorable, while only four were opposed. The newspapers' editorial position on reapportionment affected news coverage: those supporting the Court presented a more restrained account of the Court's rulings in both headlines and reportage of critical reaction to the decisions than did those papers opposed to the decision.[25]

Public Opinion Polls

Other evidence about public opinion has come from questions asked of the general public by the national commercial polling organizations, from studies by political scientists of particular groups in the population, and (to be discussed in the next section) from intensive analysis of state and national surveys carried out by university polling units. In part because the polling organizations and social scientists have examined public opinion among a variety of different populations, no unified statement about public opinion and the Supreme Court has yet been developed. Differences in the results reported in the remainder of this chapter may be somewhat confusing, but they are only an indication of our relatively underdeveloped understanding of the subject.

In the 1960s, national polling organizations began to ask how people rated the Supreme Court. Overall ratings remained fairly constant in the 1963–1967 period. A November 1966 Harris Poll showed the public giving the Court an overall negative rating (46 percent–54 percent). Younger people, the better educated, and blacks backed the Court, while southerners, older people, and the less well educated were the Court's severest critics. Similar results appear in a 1967 Gallup Poll. The Court's work was rated excellent by 15 percent, good by 30 percent, fair by 29 percent, and poor by 17 percent, leaving an almost even balance between favorable and unfavorable reactions. The June 1968 Gallup Poll showed how quickly opinion can change. Evaluations had shifted to 36 percent favorable, 53 percent unfavorable, with only 8 percent rating the Court's work as excellent. Republicans were most critical of the Court, and those with less than a college education were also negative; Democrats and those with a college education were evenly divided.

Although the Burger Court's criminal procedure rulings were more in tune with public opinion, they did not produce an improvement in the Court's overall rating. The Court received only a slightly higher Gallup Poll in 1973 than it had received in 1969—a rating much lower than the Court's mid-1960s ratings. In 1973, 37 percent rated the Court's work good or excellent, but only 6 percent rated it excellent. Thirty-five percent thought the Court too *liberal*, while 26 percent thought it too conservative. Who liked the Court had also changed. Decreased approval of the Court was shown by the college-educated, those twenty-one to twenty-nine years old (who showed the greatest drop), westerners and easterners, and Democrats, but ratings were up among those age fifty and over, southerners, and Republicans (up 12 percent).[26]

Information concerning opinion about the Supreme Court among specific groupings in the population is much more sparse, but also provides evidence of difference among groupings in views of the Court. A 1970 poll of Providence, Rhode Island, lawyers showed they were more dissatisfied with the Court than the Gallup Poll's national sample. They were also more undecided than the national sample's college-educated and business and professional groups, with their indecision coming from indifference rather than ambivalence. The criminal procedure decisions seemed most important for the lawyers who criticized the Court; they tended to hold a "traditional" attitude toward the Court—including the idea that the Court only finds law —and wanted "strict constructionist" judges. Those favoring the Warren Court were more likely to approve of judges engaging in lawmaking.[27] Another "special population," small-town police chiefs in southern Illinois and western Massachusetts, perceived changes in the Court's direction. However, some of the Illinois chiefs, who were not favorably disposed toward the Warren Court and gave the Supreme Court only a marginally positive overall evaluation in 1972, seemed unaware of changes the Burger Court produced. Massachusetts chiefs, who gave the Court a generally favorable evaluation and who were more knowledgeable about it, felt that the changes were favorable to the police.[28]

More than three of four federal judges, state supreme court judges, and lawyers surveyed in early 1977 preferred the Burger Court to the Warren Court.[29] Almost all (98.8 percent) thought the Burger Court to be more conservative than the Warren Court, and 78.4 percent approved of this conservative position. Although two-fifths thought the Court was still taking questions the respondents felt better left to the legislative or executive branches of government, 84.2 percent thought the Burger Court less likely to do so than the Warren Court, and only 2.4 percent thought the Burger Court *more* likely to do so. Over half of those in the poll approved of the Court's making it more difficult for citizens to use the federal courts to obtain redress of grievances and just under half (48.4 percent) said that the Court *was* making it more difficult. The quality of the Burger Court's written opinions was thought to be better than the quality of the Warren Court's opinions by 33.6 percent, with over half (51.9 percent) saying the quality was the same and 14.5 percent saying it was worse. Despite the perceived improvement, almost two-thirds (65.3 percent) said that the Court's opinions were often unclear and over three-fourths said they were often too long.

The national polls reported above showed not only changes in overall ratings of the Supreme Court but also shifts in confidence in the Court as an institution. In 1966 a Harris Poll majority (51 percent) expressed a great deal of confidence in the Court. This was almost 10 percent more than expressed comparable regard for either Congress or the executive branch but less than for medicine, colleges, or the military. A Gallup Poll the next year indicated that almost half the American public thought the Court had been impartial, but 30 percent—particularly older citizens, Republicans, and southerners—believed the judges showed some group favoritism.

Confidence in the Court had decreased noticeably by 1972, only 28 percent of a Harris survey expressing a great deal of confidence in it. Yet this figure was up from 1971 (21 percent) and was about the same as the confidence shown in the executive branch but more than was shown for Congress. Confidence in the Court then rose for two more years—to 33 percent having a great deal of confidence in 1973 and to 40 percent in 1974—but then decreased substantially to only 28 percent in 1975 and to 22 percent in 1976. This drop was part of the overall decrease in public confidence in its leaders, but the Court remained considerably ahead of the other two branches of government. In the 1975 survey, 38 percent said that those in charge of the Supreme Court "really know what most people they represent or serve think and want," somewhat better than Congress, the White House, or the executive branch. However, more—43 percent—said the Court was out of touch with those it served. (Nineteen percent did not know, higher than for any other institution.)

In addition to their general evaluation of the Court, the public expressed reaction to specific decisions. Thus the 1966 Harris survey showed the public, despite its overall negative rating of the Court, favoring rulings on reapportionment (76 percent–24 percent) and desegregation of schools and public accommodations (both 64 percent–36 percent). Only reapportionment received approval in the South, where only 44 percent of the public supported school desegregation. Opinion was evenly divided on the Court's having forbidden the State Department to deny passports to Communists, but the school prayer and *Miranda* rulings were disliked (30 percent–70 percent and 35 percent–65 percent). A 1973 Gallup Poll revealed that 58 percent favored the Burger Court's conservative ruling on obscenity, but roughly the same percentage disapproved of the Court's 1972 invalidation of the death penalty. (President Nixon's appointees, who had dissented, were closer to public opinion.) The rulings denying news reporters a First Amendment right to protect confidential sources and invalidating aid to parochial schools were also opposed by majorities of those questioned.

Reactions of specific groups to particular decisions have been little studied. Examination in 1967 of the impact of *Miranda* showed that four-fifths of all police officers in four Wisconsin medium-sized cities disapproved of the ruling. There was no relationship between attitude and amount of formal education, but those with least police experience approved of the decision the least.[30] Five years later, Illinois and Massachusetts small-town police chiefs did not show this same resistance to *Miranda*, as most officers had learned to live with the ruling. However, although Massachusetts officers generally favored the (*Mapp*) rule excluding illegally seized evidence from trials, it was hard to find an Illinois officer who could say anything good about that doctrine. The chiefs' views were not related to their formal education, but those who had had law enforcement training before becoming officers were somewhat more likely to view the rule positively.[31] A study of officers in twenty-nine St. Louis area police departments carried out at the same time, however, showed weak relationships between training and attitudes. Officers with a college education were somewhat less likely than others to see the Supreme Court's criminal procedure rulings as harmful, but some findings ran counter to the hypotheses that officers with higher levels of education and training would "have a view of the goals of law enforcement which includes the protection of civil liberties even of persons suspected of criminal acts" and would "be less critical of Supreme Court decisions."[32]

A 1975–1976 study of University of Tennessee Law School students showed general agreement that five major cases—*Brown*, *Miranda*, *Engel v. Vitale* (school prayer), *Roe v. Wade* (abortion), and *Furman v. Georgia* (the 1972 death penalty ruling)—were "basically good . . . , involved reasonably practical solutions to the problems presented to the Court . . . , and were appropriate decisions for the Court to be making." *Miranda* received the most positive rating on a "good-bad" dimension, *Furman* the least positive rating (marginally positive). *Furman* was also seen as the least effective of the five on the "practical-impractical" dimension, with the abortion ruling first and *Miranda* next. *Miranda* also received the highest rating on the "appropriate-inappropriate" dimension, despite criticisms that the Court had been "legislative" in its action. The respondents' responses across all three dimensions were significantly affected by the law students' legal values; those adhering to a "social welfare" view of the law were uniformly more positive about all five cases than were those adopting a more traditional "entrepreneurial" view of the law and the lawyer's role. (Political party leanings and party activism helped explain differences, but to a lesser extent.)[33] In the above-noted survey of federal and state judges and lawyers, a majority of those surveyed approved of the Burger Court's recent decisions and direction in all areas except obscenity/pornography, where 51.4 percent disapproved. Highest approval—80.4 percent—came in the area of racial discrimination. Also receiving over 70 percent approval were the Court's decisions and direction in the areas of criminal defendants' rights, sexual discrimination, labor union law, environmental protection, and free speech and freedom of the press. The Court's abortion rulings received approval from almost two-thirds (65.3 percent), and its work in the states' rights-federalism area was approved by 68.7 percent. Only 57.5 percent approved the Court's death penalty decisions, and the smallest favorable majority (51.1 percent) approved the Court's rulings with respect to mandatory busing to alleviate segregation.

More Extensive Surveys

A more detailed picture of the structure of public opinion about the Court and particularly about "public orientations toward courts [which] apparently change slowly over time if at all, as decisions and popular policy preferences (and other factors) interact,"[34] is provided by systematic opinion surveys conducted during the 1960s. These avoid the problems of the forced-choice responses of the Gallup and Harris polls, which allow people to answer without knowing about the decisions on which they are commenting. The principal extensive surveys, all undertaken during the Warren Court, were conducted in Seattle (1965), Wisconsin (1966), and Missouri (1968), and nationally by the University of Michigan Survey Research Center (SRC) as part of its 1964 and 1966 postelection surveys.

Seattle residents had relatively little information about the Supreme Court. However, they were supportive of the Court's work, although those with negative views held them somewhat more intensely. People's limited awareness appeared in the national survey through inability to name good or bad things the Court had done and through the attribution to the Court of cases it had not decided. In a test of knowledge about decisions, only 2 percent of the Wisconsin sample had all items correct and only 15 percent had more than half correct; 12 percent had every answer wrong. One-fourth of the Missouri sample declined to comment on the Court or pled ignorance. Lack of awareness in Wisconsin extended even to criminal procedure decisions which had been the subject of open controversy and to reapportionment despite the redistricting in the state. The prayer rulings were among the most salient decisions and, along with civil rights decisions, accounted for more than two-thirds of the 1964 likes and dislikes about the Court; by 1966, however, most likes and dislikes were accounted for by criminal procedure decisions. The school prayer decisions were "unknown only to the same seemingly irreducible number of persons who have managed to remain unaware of the segregation decisions."[35] A majority of blacks, however, were in such a category.[36] It thus appears that people have difficulty relating decisions to their personal lives and that "only a few cases are sufficiently dramatic to rise above the public's threshold of attention."[37] Yet because some in the Missouri survey who did not read newspapers knew of at least one of the Court's decisions, it appears that "issues which gain the Court renown (or notoriety) are so salient that they come through even to people virtually isolated from the printed word."[38]

If we move from awareness to support, we find that in the 1964/1966 SRC surveys, those giving the Court diffuse (general) support outnumbered by four to one those giving the Court specific support (that based on particular decisions). Only about 20 percent were negative as to diffuse support, but one-third were negative on specific support. Thus, despite the unpopularity of recent decisions, the Court seemed to retain a "substantial reservoir" of general support, some of which came from those opposed to particular decisions. (Among small-town police chiefs, negative evaluation of the Court did not prevent agreement by substantial majorities that the Supreme Court should be the "final judge of law enforcement and police matters.") Some of those least knowledgeable about the Court, who also seemed most trusting of government, were among those giving the Court diffuse support, while those aware of the Court were more likely to distrust government and not to find it responsive.

Higher knowledge about the Court seemed to correlate with greater *dis*-approval of the Court in Wisconsin. Similarly, in Seattle, "critics are more likely to have paid some attention to the Court than its supporters" did. Those with college degrees were more likely to be strong supporters or critics, while those with little education were likely to be weak supporters, weak critics, or neutral.[39] Favorable prior information about the Court had an independent effect on support, while favorable attitudes and belief in specific procedural rights affected what a person heard and read about the Court and reduced the effect of communication concerning the Court. In the Seattle and Wisconsin studies, but not in the Missouri survey, political party identification affected a person's attitudes. Being a Republican affected reactions to Court rulings, but being a Democrat influenced what one heard, leading to maintenance of current attitudes. In Wisconsin, although approval of the president was positively related to approval of the Court, regardless of one's party, part of Republicans' unhappiness with the Court seemed to stem from Democratic control of the White House.[40] (Republicans and Democrats also differed more than did conservatives and liberals.) In the SRC national surveys, respondents' "attitudes toward public policy as measured by the scale of liberalism/conservatism" was the single factor best able to assist in explaining support for the Court, even among those most active politically, although specific support by those knowledgeable about the Court also helped explain their general support.[41] Congruence between liberals' political positions and the Warren Court's decisions meant that liberals gave the Supreme Court more support at that time.

To what extent does the public believe in the myth that the Court is a legal rather than a political body? When the Missouri sample was asked the Supreme Court's main job, only 8.1 percent associated the Court directly with the Constitution (references to constitutionality increase the figure to 16.6 percent). Thirty percent of the responses contained references to law, and courtlike functions were mentioned by an equal number. Less than half of the comments (46.7 percent) contained positive symbols regarded as legitimators of the Court's authority, although a majority of the respondents (60 percent) did see the Court in terms of symbols and beliefs. In less than one-fourth of all the responses was the Court seen primarily as a political institution. This data led to the conclusion that "the Court's myth enjoys widespread diffusion," although members of the public were not fully dependent on external symbols to legitimate their belief in the Court.[42] However, education increased rather than reduced belief in the myth, which was not disturbed by the Court's exercise of its political role. That those of higher social status have a greater belief in the myth helps to explain its "cultural dominance"; those who do not believe in the myth were not more politically sophisticated but actually were less well socialized into prevailing norms.[43] The Wisconsin data suggests that, although myth acceptance may result from satisfaction with the Court, "mythification" increases support for the Court.

A related important question explored in analysis of the SRC national surveys was whether people thought it proper for the Court to produce changes in governmental structure or process—like those required by the reapportionment decisions—that is, whether the Court could "legitimate regime change." A Supreme Court ruling can perform this function only for those who perceive the Court, recognize that it may properly interpret and apply the Constitution, and feel that the Court was acting competently and impartially. Forty percent of the population satisfied neither of the first two conditions, while roughly one-fourth satisfied both; only about one-eighth satisfied all three. (Local leaders in Wisconsin were far more likely than the average citizen to satisfy comparable criteria.)

The Court's ability to legitimate change for only a few supports Adamany's assertion that the Court does not perform the legitimating role. However, we also need to know whether those opposed to the Court will act on their beliefs. Those who think the Court's performance is poor are "most likely to act to change a decision" but "they are neither numerous nor particularly rebelliously inclined."[44] In Wisconsin only a few of those who said they would do something to change a Supreme Court decision they disliked would try to develop further opposition to those decisions among the public. Half would work through their congressmen, and another one-fourth would "act within the established legal processes." Virtually no Illinois or Massachusetts small-town police chiefs said they would refuse to go along with a court ruling. Over half the Illinois officers said, however, that they would do something about a decision they disliked, but for most this was only to talk about the disliked decisions, complain, and "gripe." In Massachusetts, most spoke of writing to the authorities, including their commanding officers or other chiefs, in order to express their opinions, while a couple would have acted through the legal system.

These responses lead to the remark made about the Wisconsin sample: "This is quiescence indeed" and to the conclusion that "the decisions of the Supreme Court have had more effect on the reputation of the Court than the activities of its antagonists."[45] However, it is also clear that acquiescence rather than active approval has served to produce compliance with Court decisions. Because the public becomes aware of so few decisions and because public officials can often put the Court's commands into effect without involving the public, the public's views of the Court often are irrelevant to much of what the Court does. Public opinion on the Court's actions and overall estimates of the court are responses to factors external to the Court, including partisan and ideological predispositions. If the Court's doctrines become more consonant with elements of public opinion, it is likely to happen more because new justices themselves embody the changing trends in public opinion than because the Court responds directly to public opinion as registered in Harris and Gallup polls or in elections. However, the Court's strategy is to some degree affected by public opinion, as the justices try not to stray too far from that opinion and particularly from the views of its special attentive audiences.

NOTES

1. See Frank R. Strong, "President, Congress, Judiciary: One Is More Equal than the Others," *American Bar Association Journal* 60 (September 1974): 1050–52.
2. For an examination of this case, see C. Peter McGrath, *Yazoo: Law and Politics in the New Republic* (Providence, R.I.: Brown University Press, 1966).
3. See *Hylton v. United States*, 3 Dall. 171 (1796); *Calder v. Bull*, 3 Dall. 386 (1798).
4. *Little v. Barreme*, 2 Cr. 170 (1804).
5. Benjamin Cardozo, *The Nature of the Judicial Process* (New Haven: Yale University Press, 1921), p. 92.
6. Oliver Wendell Holmes, in *Collected Legal Papers* (New York: Harcourt, Brace, and Howe, 1920), p. 295.
7. For a recent discussion of the exercise of judicial review, containing a concise summary of most instances of its exercise, see Robert J. Harris, "Judicial Review: Vagaries and Varieties," *Journal of Politics* 38 (August 1976): 173–208.
8 In not every instance, however, does the Court hold that federal legislation preempts state action, for example, allowing state regulation of pollution where it would supplement rather than undercut national statutes. *Askew v. American Waterways Operators*, 411 U.S. 325 (1973).
9. Lief H. Carter, "What Courts *Should* Make Policy: An Institutional Approach, *Public Law and Public Policy*, edited by John Gardiner (New York: Praeger, 1977), pp. 141–57.
10. For an extended treatment, see Howard Dean, *Judicial Review and Democracy* (New York: Random House, 1966).
11. Reversing, respectively, *Chisholm v. Georgia*, 2 Dall. 419 (1793); *Pollock v. Farmers Loan & Trust Co.*, 157 U.S. 429 (1895); *Oregon v. Mitchell*, 400 U.S. 112 (1970); and *Dred Scott v. Sanford*, 19 How. 393 (1857).
12. Robert Dahl, *Democracy in the United States* (Chicago: Rand McNally, 1972), pp. 201–202, 207–208. The original statement was Dahl, "Decision-Making in a

Democracy: The Supreme Court as a National Policy-Maker," *Journal of Public Law* 6 (Fall 1957): 279–95.

13. Stuart Nagel, *The Legal Process from a Behavioral Perspective* (Homewood, Ill.: Dorsey, 1969), pp. 248, 259.

14. Bradley C. Canon and S. Sidney Ulmer, "The Supreme Court and Critical Elections: A Dissent," *American Political Science Review*, 70 (December 1976): 1215–18; Richard Funston, "The Supreme Court and Critical Elections," *American Political Science Review*, 69 (September 1975): 795–811.

15. David Adamany, "Legitimacy, Realigning Elections, and the Supreme Court," *Wisconsin Law Review* (1973): 825, 822.

16. Ibid., p. 845.

17. Glendon Schubert, *Constitutional Politics* (New York: Holt, Rinehart, and Winston, 1960), p. 257.

18. Harry Stumpf, "The Political Efficacy of Judicial Symbolism," *Western Political Quarterly* 19 (June 1966): 293–303, and "Congressional Response to Supreme Court Rulings: The Interaction of Law and Politics," *Journal of Public Law* 14 (1965): 376–95.

19. Jonathan D. Casper, "The Supreme Court and National Policy Making," *American Political Science Review* 70 (March 1976): 3.

20. Charles Warren, *The Supreme Court in United States History* (Boston: Little, Brown, 1922), vol. 2, p. 358.

21. Ibid., vol. 3, p. 244. That Justices Strong and Bradley, who helped produce the second decision, were named to the Court on the day of the first decision did not help matters.

22. Merle Fainsod, Lincoln Gordon, and Joseph C. Palamountain, Jr., *Government and the American Economy*, 3rd ed. (New York: W. W. Norton, 1959), pp. 540, 541.

23. Arthur M. Schlesinger, Jr., *The Politics of Upheaval* (Boston: Houghton Mifflin, 1960), p. 489.

24. Stuart Nagel and Robert Erickson, "Editorial Reaction to Supreme Court Decisions on Church and State," *Public Opinion Quarterly* 30 (Winter 1966–67): 647–55; also in Nagel, *The Legal Process From a Behavioral Perspective*, pp. 285–93. For a study of editorial reaction to a particular decision in a single state, see Hugh David Graham, *Crisis in Print: Desegregation and the Press in Tennessee* (Nashville, Tenn.: Vanderbilt University Press, 1967).

25. Chester Newland, "Press Coverage of the United States Supreme Court," *Western Political Quarterly* 17 (March 1964): 15–36.

26. See Richard Claude, "The Supreme Court Nine: Judicial Responsibility and Responsiveness," *People Vs. Government: The Responsiveness of American Institutions*, edited by Leroy N. Rieselbach (Bloomington, Ind.: Indiana University Press, 1975), particularly table 1, p. 123.

27. Edward N. Beiser, "Lawyers Judge the Warren Court," *Law & Society Review* 7 (Fall 1972): 139–49.

28. Stephen L. Wasby, *Small Town Police and the Supreme Court: Hearing the Word* (Lexington, Mass.: Lexington Books, 1976), p. 81.

29. *U.S. News & World Report* 82 (No. 9, March 7, 1966): 58, 60–67.

30. Neal Milner, *The Supreme Court and Local Law Enforcement: The impact of Miranda* (Beverly Hills, Calif.: Sage Publications, 1971).
31. Wasby, *Small Town Police and the Supreme Court*, p. 82.
32. Dennis C. Smith and Elinor Ostrom, "The Effects of Training and Education on Police Attitudes and Performance: A Preliminary Analysis," unpublished manuscript, 1973, p. 11.
33. Gregory J. Rathjen, "The Impact of Legal Education on the Beliefs, Attitudes, and Values of Law Students," *University of Tennessee Law Review* (Fall 1977).
34. Kenneth Dolbeare, "The Supreme Court and the States: From Abstract Doctrine to Local Behavioral Conformity," *The Impact of Supreme Court Decisions: Empirical Studies*, edited by Theodore H. Becker and Malcolm Feeley, 2nd ed. (New York: Oxford University Press, 1973), p. 203.
35. Kenneth Dolbeare, "The Public Views the Supreme Court," *Law, Politics and the Federal Courts*, edited by Herbert Jacob (Boston: Little, Brown, 1967), p. 199.
36. Walter Murphy and Joseph Tanenhaus, "Public Opinion and the United States Supreme Court: Mapping of Some Prerequisites for Court Legitimation of Regime Change," *Frontiers of Judicial Research*, edited by Joel B. Grossman and Joseph Tanenhaus (New York: John Wiley, 1969), p. 284.
37. John H. Kessel, "Public Perceptions of the Supreme Court," *Midwest Journal of Political Science* 10 (1966): 175.
38. Gregory Casey, "The Supreme Court and Myth: An Empirical Investigation," *Law & Society Review* 8 (Spring 1974): 397.
39. Kessel, "Public Perceptions of the Supreme Court," pp. 187–88.
40. Kenneth M. Dolbeare and Phillip E. Hammond, "The Political Party Basis of Attitudes Toward the Supreme Court," *Public Opinion Quarterly* 31 (Spring 1967): 23–24.
41. Walter F. Murphy, Joseph Tanenhaus, and Daniel L. Kastner, "Public Evaluation of Constitutional Courts: Alternative Explanations," Sage Professional Papers #01–045 (Beverly Hills, Calif.: Sage Publications, 1973), p. 50.
42. Casey, "The Supreme Court and Myth," p. 398.
43. Ibid., p. 402.
44. Dolbeare, "The Public Views the Supreme Court," p. 208.
45. Kessel, "Public Perceptions of the Supreme Court," p. 191.

PART TWO | THE STRUCTURE OF THE COURTS AND JUDICIAL SELECTION

The United States has a dual court system. There is a set of national courts, including the territorial courts, and the states have their own courts. This gives us fifty-two separate court systems—the national one, one for each of the fifty states, and a separate set of courts for the District of Columbia. The Supreme Court is the highest court for all those court systems. The national court system has four levels and some specialized courts. The Supreme Court is the only court specifically designated in the Constitution. In addition, there are the courts of appeals, which are intermediate-level appellate courts; the basic trial courts, called district courts; and the U.S. magistrates, until recently U.S. commissioners.

The national court system's four levels are similar to the basic levels of most state court systems. In addition to specialized courts, for example, for probate (wills and estates), juvenile proceedings, or small claims, each state has general jurisdiction trial courts, called district courts or circuit courts, which may hear all civil and criminal cases; an intermediate-level appellate court, to which all or most cases are first appealed; and a supreme court (in some states called the court of appeals). In states with no intermediate-level appellate courts, the supreme court handles all appeals. Where an intermediate court has been added to assist in handling caseload, supreme courts tend to hear less private litigation and more constitutional and criminal litigation, particularly the latter.* Some states retain a fourth level of courts eliminated elsewhere—the so-called inferior (meaning lower) or "petty" courts. These include justice-of-the-peace courts, mayor's courts, police courts, and some special urban courts established to assist with the higher volume of litigation in large cities. These limited-jurisdiction courts often do not maintain a transcript of proceedings and thus are not "courts of record." As a result, appeals from their decisions usually go to the general jurisdiction trial courts for a new trial (trial *de novo*) instead of directly to an appellate court.

Although the Supreme Court is our principal interest, the other federal courts must receive our attention. The work of the federal district courts and courts of appeals is crucial for the Supreme Court. If cases were not filed in the district courts and then appealed to the courts of appeals, there would be little business to take to the Supreme Court. On the other hand, if many cases, once initiated, are appealed all the way to the Supreme Court, it may become overloaded, although once presented with a case, the Supreme Court may refuse to review it or may dispose of it summarily instead of giving it "full-dress" treatment. The flow of cases gives the Supreme Court much discretion, but because it is not able to initiate cases, it must rely on what is brought in the lower courts and on the way cases are framed there. The district courts and courts of appeals are also crucial for the system because most litigation terminates there, making those courts more important than the Supreme Court for most of those who have contact with the judiciary. Moreover, they make much law in the areas of litigation not reviewed—or not reviewed often—by the Supreme Court.

Although the next part of the book concentrates on the Supreme Court, this part is intended to provide a picture of the federal court system's overall structure and of the selection of judges for all the courts of that system. In Chapter 4 we look at the four basic levels of the national court system and at its specialized courts as well as at the federal regulatory agencies, which perform an important judicial role. Discussion of some of the judicial system's personnel is also included, as is an examination of judicial administration, increasingly important because of the courts' increasing caseload and the attention Chief Justice Burger has focused on the subject. Chapter 5, which draws primarily on the last few presidential administrations, provides a discussion of the process by which Supreme Court nominees are confirmed, and of the background and qualifications of federal judges.

*See Burton M. Atkins and Henry R. Glick, "Environmental and Structural Variables as Determinants of Issues in State Courts of Last Resort," *American Journal of Political Science* 20 (February 1976): 113.

3 | THE SUPREME COURT AND THE FEDERAL JUDICIAL SYSTEM

THE FEDERAL COURT SYSTEM: BASIC ELEMENTS

The Supreme Court

We are so accustomed to a United States Supreme Court of nine justices that it would be difficult to alter the number. However, while the Constitution provides that there be a Supreme Court, its size is not designated there. The Court at first had fewer than nine justices, and until President Grant's time the number fluctuated. This resulted from presidents' leaving seats vacant, as Lincoln did with three positions, and from Congress's ordering that positions not be filled, as it did during Reconstruction, or increasing the number of seats—to ten during the Civil War. Most recently, President Franklin Roosevelt would have added a justice for each one over the age of seventy who did not retire. However, Chief Justice Hughes's statement to Congress that a court of more than nine judges would be unwieldy and that because the Constitution speaks of "one Supreme Court," the justices could not divide into panels or "divisions" to hear cases as some state high courts do, helped defeat the "Court-packing" plan.

The Court's quorum for doing business is six justices. Thus it can operate during vacancies, illnesses, and recusance (withdrawal by a justice from participation in a case). However, an evenly divided court (4–4 or 3–3) means that the lower court's ruling is affirmed without the justices writing an opinion. In order to avoid criticism, even from within the Court, the Court has generally tried to wait until it is fully staffed before deciding controversial cases. Recently, however, faced with a challenge to statutes providing for summary repossession of property, the Court, with two vacancies (those filled by Justices Powell and Rehnquist), failed to delay deciding the case until the new justices were seated and could participate in reargument. The result was a 4-to-3 vote followed shortly by the full Court's substantial modification of that ruling. The resultant indecision was said by Justice Blackmun to have been similar to the confusion surrounding the Legal Tender Cases and to have created an unfavorable image of a "wavering tribunal off in Washington, D.C."[1]

If it is acceptable to wait for a vacancy to be filled in order to minimize division in the Court, participation by a justice in a case merely to create a full Court is not necessarily proper. Usually justices recuse—without explanation or comment—because of some involvement with a case before they became justices or because of their acquaintance with the parties or the lawyers. However, shortly after he joined the Court, Justice Rehnquist precipitated controversy by participating in an important challenge to Army surveillance of civilian political activity. He did this although as assistant attorney general he had testified before Congress on military surveillance and had commented that the case, already initiated, was nonjusticiable, that is, not appropriate for judges to decide—the central procedural issue in the case. The Supreme Court's vote without his participation would have been 4–4, sustaining the lower court ruling that the case could be heard, but Rehnquist cast the crucial fifth vote to reverse the lower court. This caused the American Civil Liberties Union (ACLU) to petition for a rehearing of the case in which Rehnquist's disqualification was specifically requested. The Court denied the hearing petition without comment—its usual procedure. Rehnquist, however, wrote a memorandum supporting his participation. He said he had only been the government's attorney, expressing the administration's position, not necessarily his own. He added that someone in government service prior to joining the Court was quite likely to have made statements of opinions on subjects which would arise later in litigation. To have a blank mind on such subjects would, he said, "be evidence of lack of qualification, not lack of bias." The American Bar Association code of judicial ethics showed a clear preference for avoiding even the appearance of impropriety, but Rehnquist stressed a duty to participate, particularly where his vote was necessary to resolve the case.[2]

Only the Supreme Court's original jurisdiction—cases brought to it directly—is detailed in the Constitution, while the Court's appellate jurisdiction is open-ended, subject to modification by Congress. (See Article III, Sec. 2.) From time to time, Congress has altered the federal courts' overall jurisdiction, and Congress once removed some of the Supreme Court's appellate jurisdiction. This successful effort came in the aftermath of the Civil War when the Reconstruction Congress was at odds with President Andrew Johnson and was fearful of the Court's potential action concerning questionable detentions of citizens. A law was passed removing the Court's jurisdiction over cases arising under the Habeas Corpus Act. The Court, although it had already heard argument in a case under the statute, *Ex parte McCardle*, dismissed the case, saying it no longer had the power to hear it. There have also been a number of recent attempts to limit the appellate jurisdiction further, usually the result of legislators' unhappiness with the Court's decisions, for example, on reapportionment. None of these attempts have been successful, although the Jenner-Butler bill to strip the Court of its power to hear certain internal security cases failed in the Senate by only one vote.[3]

The Supreme Court's original jurisdiction accounts for only a very small proportion of its workload; there have been only approximately 150 such cases decided in the Court's history. The Court said in *Marbury v. Madison* that original jurisdiction was limited to the subjects so designated in Article III of the Constitution. The Court has also made clear that it will change its original jurisdiction into appellate jurisdiction by ruling that most such cases can first be heard in the trial court, only later coming to the Supreme Court for review. For example, when Ohio tried to file an original jurisdiction case against several companies for mercury pollution of Lake Erie, the Supreme Court ruled that state courts in Ohio could deal with the problem in terms of local law. The justices also said that Illinois's suit against Milwaukee for polluting Lake Michigan could be tried on nuisance law in the lower federal courts. Because other federal courts could hear the case, the Court also refused to accept in its original jurisdiction an antitrust case brought by several states against auto manufacturers for conspiring not to develop pollution-control equipment.[4]

Such rulings leave only cases between two states—for example, over boundaries or water allocation—or between a state and the national government for the Court's original jurisdiction. The suits by states to test the validity of the 1965 Voting Rights Act (*South Carolina v. Katzenbach*) and its 1970 successor (*Oregon v. Mitchell*) are examples of the latter. Yet even when two states are involved, the Court may try to send cases elsewhere. When Arizona sued New Mexico in the Supreme Court over the latter's tax on the generation of electrical energy, said to affect Arizona citizens and to burden interstate commerce, the Court ruled that all interests would be protected in a suit by the affected utilities in New Mexico courts, with appeal to the Supreme Court available.[5]

When the justices do accept an original jurisdiction case, they appoint a special master (often a senior federal judge) to hear testimony and to make findings and recommendations; the master really serves as the equivalent of a trial judge. After lawyers' briefs and argument, the justices decide whether to accept, modify, or reject the master's recommendation, acting, as usual, in what amounts to an appellate capacity.[6]

The bulk of the Supreme Court's work is in its appellate jurisdiction, in which cases come to the Court in basically two ways—*appeal* and *certiorari.* (Quite infrequently, the Court receives a case on *certification.* Here the lower court, faced with a new legal question, before resolving the case "certifies" the question for answer by the Supreme Court.) The Supreme Court long had to hear all cases appealed to it; they came to the Court on a *writ of error.* After the present courts of appeals were established, the Supreme Court was provided with its certiorari jurisdiction, that is, the authority to pick and choose the cases it would hear, which accounts for the great majority of cases. The Court's appeals jurisdiction encompasses the remainder. In theory the Court must hear cases in the appeals jurisdiction, but the Court has made this mandatory jurisdiction discretionary, so that it is similar in many ways to the formally discretionary certiorari jurisdiction. (See pp. 143–53 for more extended discussion.)

In addition to its basic task of deciding cases appealed to it, the Supreme Court has a potentially important role to play in the development of procedural rules for the federal judiciary. Some of these are announced in the course of deciding specific cases and are said to come from the Court's "supervisory power" over the lower federal courts. The "*Mallory* rule" (from *Mallory v. U.S.*) requiring prompt arraignment of a defendant in a federal criminal case is one example. In another, *U.S. v. Hale,* the Court held that a defendant's silence—in not immediately reporting an alibi to the police—was not proof of his guilt. (Because they are based on the supervisory power rather than a specific provision of the Bill of Rights, such rules are only application to federal, not state, court proceedings.)

Rules are developed only occasionally through cases. More complete sets of rules—the Federal Rules of Criminal Procedure, Civil Procedure, Evidence, Appellate Procedure, and Bankruptcy—are promulgated by the Court under a grant of authority from Congress. The Court does not itself develop the rules; an advisory committee or, more recently, a committee of the Judicial Conference prepares them, the Conference adopts them, and the Supreme Court then issues them. The rules then go into effect unless disapproved by Congress, usually within ninety days. This procedure was often criticized by Justices Black and Douglas. They felt that the Court should restrict itself to deciding cases, and that because the Court could not give full consideration to the Conference-proposed rules, the Conference itself should have full power to promulgate them. The procedure has also been described as a

> unique situation of rules drafted by a committee of private citizens and judges acting in an advisory capacity, which operates for the most part in private; approved by a body of judges, meeting entirely in private; promulgated by the Supreme Court without any real expectation. or the procedure to warrant that

expectation, of focused consideration of constitutional or statutory questions; and "approved" by the legislature through simple inaction for a period of ninety days.[7]

For years, Congress accepted the Court-announced new procedural rules without debate. However, when the Supreme Court announced new Federal Rules of Evidence in 1972, Congress, because of substantive disagreement over a number of the rules which embodied noticeable change from the past, delayed the effective date, subjected the rules to section-by-section reconsideration, and enacted them only in 1975. At the same time, the ninety-day "waiting period" for congressional consideration of the rules was doubled; each house of Congress could extend that period for a specified time or until the rules were specifically approved by legislation. Congress also postponed for a year the effective date for the Federal Rules of Criminal Procedure, a further indication that in the future the Court would have less freedom to make the rules for the federal judiciary.[8]

The Courts of Appeals

Until 1891, there was no intermediate-level appellate court in the national court system. The membership of the earliest circuit courts consisted of two Supreme Court justices ("circuit justices") and one district judge, with the latter doing most of the work. In 1869 Congress provided a circuit judge for each circuit to relieve the Supreme Court justices of their "circuit riding." In 1891 the Courts of Appeals Act added regular circuit judges and confirmed the intermediate appellate nature of the courts, although the old circuit courts were not dismantled for another twenty years.

Each Supreme Court justice still has responsibility for one or more circuits, particularly for emergency matters arising when the Supreme Court is not in session. Usually the Circuit Justice passes requests along to his colleagues, but he can release someone on bail and stay orders of the lower courts until the full Supreme Court has a chance to act on petitions, as Justice Powell did in the summer of 1976 in staying the carrying out of the death penalty until the full Court could rule on a petition for rehearing of its decision sustaining the penalty's constitutionality. The Circuit Justice takes action only when an applicant shows that the lower court's decision is erroneous and that irreparable injury would occur without a stay order and when the justice also decides that the other justices would consider the matter of sufficient seriousness to grant review, because the action of "the lower court, which has considered the matter at length and close at hand," is presumed to be correct.[9] (The Circuit Justice occasionally consults with the other justices who are available, although the formal action is his alone.)

For many years, there have been eleven regular United States Courts of Appeals, earlier called the Circuit Courts of Appeals. There is also a Temporary National Emergency Court of Appeals for cases arising under wage-price control and emergency natural gas programs, like the Emergency Court of Appeals established during World War II for wage-price cases, and a special court to handle problems under the Regional Rail Reorganization act of 1973; both it and the Temporary National Emergency Court of Appeals are served by judges from the regular court of appeals. The circuits covered by the regular courts of appeals, except that for the District of Columbia, cover several states and vary considerably in size. In 1976, before new judgeships were created, the First (Massachusetts, New Hampshire, Maine, Rhode Island and Puerto Rico) had only three judges, while the Ninth, the largest geographically (Alaska, Hawaii, Guam, the West Coast states, Idaho, Montana, Nevada, and Arizona) had thirteen and the Fifth (Texas through Florida) has fifteen active ("regular") judges. In 1977, Congress, acted on earlier proposals, to divide the Fifth Circuit into two new circuits and deferred acting on recommendations for dividing the Ninth Circuit (perhaps into "divisions") at least until after the naming of the last of the new judges it was to receive.

As of early 1977, there were ninety-seven authorized judgeships for all the courts of appeals; the appeals court judges are assisted by senior (semiretired) judges, district judges who sit temporarily as appeals judges "by designation," and judges from other circuits. In 1972 the Judicial Conference had asked for thirteen more circuit judgeships. Its request, expanded in 1976 and again in 1977, was for thirty-five additional judges, with ten to be for the Ninth Circuit alone. Congress's extended delay in acting on the 1972 request and a companion request for additional district court judgeships prompted Chief Justice Burger to criticize Congress in 1976. However, the delay then probably resulted from the Democratic Congress's unwillingness to give a Republican president power to name a large number of judges without Congress having a "share of the action" and thus could not be resolved easily. With a Democratic president, Congress is expected to act to pass a judgeship bill, just as it did in 1961 after President John Kennedy assumed office.

Prior to 1875, most cases involving federal questions were decided in the state courts because federal trial courts had no general "federal question" jurisdiction, no appeals were allowed in federal criminal cases, and no appeals were permitted in civil cases unless the amount at issue was over $5,000. After 1875, with these barriers removed and with regulatory legislation growing at the national level, the appeals court caseload began to grow. Particularly after the Supreme Court received the authority to handle most cases through its certiorari jurisdiction, the courts of appeals became the primary courts overseeing the development of national law. They must decide all the cases brought to them from the district courts. Thus they handle cases on all types of federal law, "traditional" areas of litigation such as admiralty, bankruptcy, antitrust, and patents as well as the constitutional issues which tend to be more prominent in the Supreme Court's work. Of particular importance are cases appealed from the federal regulatory agencies, which go directly from the agencies to the courts of appeals.

The recent growth in appeals court caseload has been phenomenal and disproportionate to growth in district court caseload. The courts of appeals terminated over 3,000 cases in 1944 but over 15,400 cases thirty years later. Roughly 10 percent of those were from the regulatory agencies, the rest from the district courts. Between 1960 and 1975 alone the appeals courts experienced a 321 percent increase in cases, while the number of judges increased only 43 percent. From 1970 to 1975, with no new judges added, case filings increased 60 percent. From 1975 to 1976, filings increased another 10 percent. Although more cases were terminated than during the previous year, the number fell short of the new filings, leaving an increased number of pending cases.

The caseload growth resulted in part from more prompt disposition of civil trials, litigants using the time gained to appeal instead of settling cases, and from the availability of lawyers for indigents to carry their criminal cases on appeal. Criminal cases now constitute about one-third of the appeals court caseload and account for the greatest growth in recent years. In Fiscal Year 1966, 46 percent of federal criminal cases in which the defendant was convicted after trial were appealed; in FY 73, the figure was 75 percent. Many of the criminal cases, particularly those involving prisoners, are considered "frivolous," that is, having little substance or not raising new issues. Many others are routine, and still other appeals are "ritualistic." Like many of the 1960s civil rights cases, they are brought because of the litigant's demands even when the likelihood of reversal of the trial court is low.[10] Only a relatively small proportion of cases involve "nonconsensual" appeals "which raise major questions of public policy and upon which there is considerable disagreement."[11]

Among the nonconsensual appeals may be those in which "issue transformation" has taken place in the appeals court, such as cases in which civil liberties issues, while not an important part of the case at trial, are central at the appeal stage. Such issue transformation was found in civil liberties cases in the Third, Fifth, and Eighth Circuits from 1956 through 1961, particularly in race relations cases. By being more favorable to blacks' claims than the district courts had been and by granting liberties denied in the district courts, the appeals courts enforced a more "national" and less local perspective.[12] However, an examination of all cases in the Second, Fifth, and District of Columbia Circuits in Fiscal Years 1965–1967—from the district courts up—led to the conclusion that there was very little issue transformation. Only 6.4 percent of the cases provided an indication that the two levels of judges defined the issues in cases differently, although "circuit judges [did] filter issues on their way to the Supreme Court."[13]

The finality of courts of appeals' decisions is particularly significant. Because very few cases are appealed to the Supreme Court and the Supreme Court rejects most certiorari petitions, in the great bulk of cases appeals court rulings are left as the final judicial statement. For example, in the Second, Fifth, and District of Columbia Circuits, only one in five decisions was appealed, with the Supreme Court granting review to only one-tenth of those—leaving over 98 percent of appeals court rulings final. Overall, certiorari petitions in court of appeals cases were granted at the rate of 6.3 percent in 1974, compared to 15.1 percent thirty years earlier and 11.8 percent twenty years earlier. While review had been granted to 3.1 percent of the terminations in the courts of appeals in 1944, the figure for 1974 was only one percent.[14] Because many of their actions are not disturbed, the appeals courts "*make* national law," although "residually and regionally," and the Supreme Court remains largely dependent on the courts of appeals "to enforce the supremacy and uniformity of national law,"[15] particularly in those areas of law in which undisturbed cases tend to cluster. This has included such important areas as workmen's compensation and minimum wage, social security, insurance contracts, and even school desegregation.

Most cases in the U.S. courts of appeals are heard by rotating panels of three judges. In some circuits, the panels sit either only at circuit headquarters (the First at Boston, the Second at New York City, the Seventh at Chicago); in other circuits, they sit in several cities (the Eighth in St. Louis and St. Paul, the Ninth in San Franscisco, Los Angeles, Portland, Seattle, Hawaii, and Alaska). Cases may also be heard by the entire court, sitting *en banc*, either initially or after a panel has heard them. Such proceedings are relatively new, not appearing in the Judicial Code until 1948. They were developed when the Supreme Court gave the appeals courts the responsibility for producing uniformity in legal matters within a circuit. Because bringing all members of an appeals court together prevents the judges from hearing cases in panels, thus slowing down the court's work, and may increase rather than decrease conflict, *en banc* sittings are not used frequently. The judges keep up with their caseload by curtailing and even eliminating oral argument in routine cases and by deciding those and other cases summarily, without opinions or with only one- or two-line orders.[16]

The appeals courts affirm the district courts at a high rate, even if one excludes the routine cases, and concurring and dissenting opinions are infrequent. The district courts thus follow appropriate legal doctrine not as a result of reversal by their "superiors" but through "informal controls," including precedent, professional socialization including service with the appeals court and attendance at meetings with its judges, anticipation of what the appeals judges will do—so that they will not be reversed frequently—and ideological unity provided by their shared regional and political backgrounds.[17]

The problems generated by increased caseload in both the Supreme Court and courts of appeals have led to proposals to restructure the national judicial system. The first major proposal was recommended by a study group appointed by Chief Justice Burger and chaired by Harvard Law School Professor Paul Freund. It called for a National Court of Appeals, a "mini-Supreme Court," staffed by present appeals court judges on a rotating basis. The new court was to sort through the cases awaiting review by the Supreme Court and select roughly four hundred each year most worthy of the Supreme Court's attention. From these, the Supreme Court was to select 120 to 150 cases to which it would give full-dress treatment. The new court's denials of review and its determinations of legal conflicts between the courts of appeals would have been final.[18] This hotly contested proposal was shelved, largely because it would have deprived the Supreme Court of considerable power through its finality of denials of review.

Attention to the need for uniform national law continued, however. In 1975 the Commission on Revision of the Federal Court Appellate System, chaired by Senator Roman Hruska (R-Neb.) and Circuit Judge J. Edward Lumbard, after study of our national appellate courts' decisional capacity and procedures, proposed a National Court of Appeals of several judges to be appointed by the president, and the proposal is presently before Congress for action. In attempting to assist the Supreme Court and to increase the national court system's capacity for "declaring and defining the national law" without increasing the diversity of legal interpretations, the commission felt that additional judges, new circuits, a changed basis of federal court jurisdiction, and new methods for disposing of caseloads would be insufficient. As proposed by the commission, the new National Court of Appeals would receive cases in two ways. The Supreme Court could send cases before deciding them ("reference jurisdiction") and the courts of appeals would send them when a rule of federal law was applicable to recurring factual situations or when federal courts had reached inconsistent decisions on a rule of federal law ("transfer jurisdiction"). The new court could refuse to accept cases, a decision which would not be reviewable, but its decisions on the merits could be reviewed by the Supreme Court on certiorari.

The District Courts

Each federal judicial district served by one of the nation's ninety-four district courts is located in a single state or territory, except for the District of Wyoming, which includes the Montana and Idaho portions of Yellowstone National Park. Twenty-six states, the District of Columbia, and the nation's four territories each constitute one district; the remaining states have from two to four districts. The district courts, which sit at statutorily designated locations, have from one to 27 judges, the Southern District of New York having the most. Prior to 1970, there were 343 authorized district judgeships. Because the energies of the district judges added in 1961 and 1966 were absorbed by increases in prisoner petitions and new federal criminal cases, there are now 404 district judge positions and Congress is likely to approve Judicial Conference requests for roughly one hundred more. All district judges are appointed by the president with the advice and consent of the Senate. The judges of the territorial courts in the Canal Zone, Guam, and the Virgin Islands, which were established under Congress's power to regulate the territories and which hear cases under local as well as federal laws, have terms of from four to eight years. All other district judges, including those in Puerto Rico since 1966, have appointments "during good behavior," that is, for life.

The District of Columbia is governed both under laws covering the entire nation and those, passed by its governing body and by Congress, applicable specifically to the District. In addition to the usual national courts—a district court and a court of appeals—the District has another set of courts, closer in function to state courts. Prior to 1971, these other courts, established under Congress's authority to make laws for the nation's capital (Art. I, Sec. 8, cl. 17), were the District of Columbia Court of General Sessions (the trial court) and the District of Columbia Court of Appeals. Cases from the latter court went not directly to the U.S. Supreme Court, as they would from the highest state courts, but to the Court of Appeals for the District of Columbia. In 1970 a new District of Columbia Superior Court assumed most local jurisdiction, including some cases formerly in the U.S. District Court. Most cases from the District of Columbia Court of Appeals now go directly to the U.S. Supreme Court.[19] Although the District of Columbia is not a state, the Supreme Court has ruled that a criminal defendant there is not entitled to be tried by an Article III (lifetime) judge. Pointing to state, territorial, and military courts, all of which enforce federal law and which do not have Article III status, the Court said that not every judicial proceeding involving an act of Congress must be presided over by an Article III judge.[20]

District court cases are usually heard by single judges. However, in special situations, three-judge district courts, composed of two district judges and a court of appeals judge, have been convened. These courts were first established in 1903 to deal with requests for injunctions against Interstate Commerce Commission (ICC) orders. The jurisdiction of these courts was extended in 1910 to cases involving state laws when it was felt that allowing single district judges to invalidate state laws gave them too much power, particularly over state economic regulation. Their jurisdiction was further extended in 1913 to requests for injunctions against state administrative actions and to federal statutes in 1937. Appeals from decisions of three-judge district courts go directly to the Supreme Court—in part because of the importance of the cases, in part because an appeals court judge has already participated in hearing them. In the 1950s and 1960s, injunctions against state laws, once sought primarily to block economic legislation, were sought against laws allegedly interfering with civil rights. From the mid-1960s through the early 1970s, the use of three-judge courts increased rapidly, bringing more cases into the Supreme Court's mandatory jurisdiction without benefit of treatment by the courts of appeals; in the 1972 Term, such cases accounted for nearly one-fourth of the Court's opinions. Under Chief Justice Burger, the Court developed doctrines to limit their use, with Justice Stewart stressing that Congress had established three-judge courts *only* to prevent state laws from being struck down by a single judge on constitutional grounds, with "the overriding policy . . . of minimizing the mandatory docket of this Court in the interests of sound judicial administration."[21] Stewart also thought that, while most appeals from three-judge courts were disposed of summarily, these mandatory jurisdiction cases hindered the Court in choosing the cases to consider fully and thus ran counter to the purpose of giving the Court its certiorari (discretionary) jurisdiction. The Chief Justice also called upon Congress to eliminate three-judge courts as well as other direct appeals to the Supreme Court. Such appeals were first eliminated with respect to ICC orders and government civil antitrust cases. Then in 1976 a law was passed to eliminate three-judge courts except in congressional and state legislative reapportionment cases or unless specifically provided for in congressional statutes like the Civil Rights Act of 1964 and the Voting Rights Act of 1976. Henceforth, requests for injunctions against state laws would be handled by single judges; even when three-judge courts were required, a single judge was to handle all preliminary matters.

The district courts are the federal system's general jurisdiction trial courts insofar as they deal with all federal cases for which jurisdiction is provided. Their caseload—now over 100,000 new civil cases and roughly 40,000 new criminal cases each year, with more than 10,000 civil and 7,500 criminal cases tried—is highly diverse. Among the criminal cases initiated by the federal government, which have increased recently, the largest number (almost one-fourth) have involved narcotics, followed by forgery and counterfeiting; there have also been substantial numbers of larceny and theft, fraud, and Selective Service cases.[22] It has been suggested that the increase in the number of criminal cases results not only directly from increases in crime but also from increased public pressures on prosecutors to increase their activities in the hope that such increased prosecution will reduce the effects of increased population growth—of which one cost is increased crime.[23]

Roughly half the civil cases brought in district court in Fiscal Year 1973 —a typical recent year—came under particular federal statutes such as antitrust, bankruptcy, commerce, labor, copyright and patent, securities, Social Security, and tax; one-fifth were contract actions; and under one-fourth were tort suits. The civil caseload also included over 17,000 prisoner petitions, which increased in recent years as a result of the Supreme Court's 1960s criminal procedure decisions and easier access to federal habeas corpus relief. However, petitions challenging convictions have begun to be replaced by those raising prison confinement issues, although those from federal prisoners are, in turn, decreasing as greater use is made of new grievance mechanisms in the prisons. Most federal district court civil cases involve private individuals or businesses raising federal questions or bringing suit under provisions which allow a citizen of one state to sue a citizen of another (diversity of citizenship); the United States is a party in only one-fourth of the civil cases.

In the late 1960s, as district courts disposed of their backlogs, more civil cases went to trial. This factor, along with increased case filings, Congress's refusal to add new judges, and Congress's stating that cases should be given "priority" or should be "expedited"—without any ranking of priorities—has put a strain on the capacity of busier district courts to keep up with their dockets. However, the Supreme Court had to point out in 1976 that district judges may *not* use their workload as a reason for refusing to accept cases properly brought to them. In overruling a district judge who had remanded a diversity-of-citizenship automobile accident case to the state court because of his crowded docket, Justice White took a poke at Congress, saying it was unfortunate "if the judicial manpower provided by Congress in any district is insufficient to try with reasonable promptness the cases properly filed in or removed to that court."[24] Congress in 1977 was, however, considering eliminating the federal courts' diversity-of-citizenship jurisdiction, a proposal strongly advocated by Chief Justice Burger.

Magistrates

The district court judges appoint referees in bankruptcy and U.S. magistrates. Bankruptcy referees (called bankruptcy judges in some areas) serve two-year terms; they admit people to bankruptcy and perform tasks related to administering the assets of bankrupts. Prior to 1946, they were paid by fees they collected, but now they are salaried and an increasing number serve on a full-time rather than a part-time basis. They handle close to 250,000 petitions from businesses and individuals each year. Proposals are now pending in Congress to provide greater independence for bankruptcy judges, for example, to make them Article III (lifetime) judges at the same level as the district courts, although other arrangements are being considered.

The position of U.S. magistrate resulted from the recent reorganization of the U.S. commissioner system.[25] The commissioners' authority was largely restricted to trying petty criminal offenses (no more than six months in jail, a $500 fine, or both) committed in federal enclaves such as military bases and national parks, where they were often the only federal judicial "presence." Commissioners could also issue search and arrest warrants, conduct preliminary hearings and bail proceedings, and file criminal complaints in more serious cases. Despite these tasks, the commissioners often did not have a clear idea of their authority or functions, and there was considerable variation from district to district in how commissioners were to deal with basic problems. The commissioners, about one-third of whom were not lawyers, were paid by a fee system with a $10,500 limit, quickly reached by the busiest commissioners.

Under the Federal Magistrates Act, passed in 1968 and in effect in all district courts in 1971, the number of full-time positions has been increased. All are to be lawyers except for some part-time magistrates serving under special circumstances. Full-time magistrates serve for eight years, part-time ones for four years. In 1976 there were 145 full-time and 350 part-time magistrates and 18 combination positions (clerks of court also serving as magistrates)—a change from the initial 80 full-time, 442 part-time, and 15 combination positions. Over two-thirds of the full-time magistrates serve in the nation's twenty-five largest federal district courts—assigned there by the U.S. Judicial Conference.

Magistrates have the commissioners' former duties, but in order to relieve district judges of some caseload, their criminal jurisdiction can be increased to include minor offenses (a maximum fine of $1,000, a maximum prison sentence of a year, or both). Thus they can now hear immigration cases, most theft cases, and serious traffic cases arising in federal jurisdiction. They also conduct preliminary review of applications for posttrial relief and Narcotics Addict Rehabilitation Act civil petitions, can conduct civil and criminal pretrial and discovery proceedings, can act as special masters in civil cases, and can review administrative records in Social Security cases. With the consent of both parties, they can conduct civil trials. In such cases, appeal is to the district court, with trial de novo to protect the individual's right to be tried by an Article III judge. However, the district judge usually bases his ruling on the magistrate's report. (Congress is likely to make further additions to the magistrates' authority in the near future.)

Despite the courts of appeals' restrictive views of the magistrates' duties, the "additional duties" assigned by the district judges have become an increasing proportion of magistrates' work each year and the traditional commissioners' tasks a decreasing proportion. In two decisions on the scope of magistrates' duties, the Supreme Court limited their authority in one case and expanded it in the other. On the basis of what it saw as Congress's intent, the Court refused in *Wingo v. Wedding* to allow a full-time magistrate to conduct an evidentiary hearing on a federal habeas corpus petition even if the hearing was recorded on sound equipment for review by the district judge. Despite Chief Justice Burger's dissenting argument that the district judge could reject or amend the magistrate's recommendation and could decide to hold a hearing himself if he wished so that the magistrate's decision would in no event be final, the Court said that submitting reports and recommendations was the most a magistrate could do. However, the Court upheld district judges' referrals to magistrates of challenges to Social Security (including Medicare) benefit entitlement determinations as "substantially assist[ing] the district judge in the performance of his judicial function, and benefit[ing] both him and the parties." In *Mathews v. Weber,* the Chief Justice pointed out for the Court that the district judge had the opportunity to object to the magistrate's findings of fact and conclusions of law and could hear the matter *de novo* if he wished.

Specialized Courts and Regulatory Agencies

In addition to the district courts and courts of appeals, there are also some specialized federal courts. The Court of Claims, the Customs Court, and the Court of Customs and Patent Appeals were once Article I courts whose judges had limited terms, but in 1948 they were made Article III courts with life-time judges.[26] The Court of Military Appeals remains an Article I court, and the Tax Court, connected with the Internal Revenue Service, is more an administrative appeals body than it is a court. The Commerce Court, no longer in existence, was established in 1910 and "disestablished" in 1913.

The Court of Claims, established in 1855 so that the United States government could be sued, has original jurisdiction over contractual, tax, injury, and other nontort claims against the United States. Cases over public contracts are the biggest part of its work. When first established, the only claims it could hear were those referred by the House of Representatives or the Senate and it did not have authority to award judgments against the government; the most it could do was report its findings to Congress along with a proposed bill authorizing payment. In the 1860s, however, its judgments were made final. The Court of Claims also has limited concurrent appellate jurisdiction with the courts of appeals over tort actions initiated against the federal government in the district courts. The court has eleven commissioners, the equivalent of trial judges, and five judges, who sit *en banc* in Washington, D.C., to review the commissioners' determinations.[27]

The Customs Court, established in 1926, has appellate jurisdiction over rulings and appraisals on imported goods made by the collectors of customs. It has nine judges, of whom no more than five may be from one political party. Although some limited direct review from its decisions is available in the Supreme Court, most of its rulings are appealable to the Court of Customs and Patent Appeals. Created earlier (1910), that court's five judges have appellate jurisdiction on matters of law over not only the Customs Court but also the Patent Office and the Tariff Commission.[28]

The Court of Military Appeals is the newest specialized court (1950). Its three civilian judges, of whom no more than two may be of one party, serve for staggered fifteen-year terms. In applying the revised Uniform Code of Military Justice (UCMJ), they must review courts-martial decisions involving general or flag officers, capital punishment decisions affirmed by the Boards of Review in the military services, and cases the service judge advocate generals (JAG) certify for review. Its principal business, however, is discretionary review of court-martial decisions involving bad-conduct discharges and prison sentences of more than one year. It considers such cases if a Board of Review approves a petition by the convicted person, granting approximately 10 percent of the petitions and denying the others without explanation. The Supreme Court may review the court's decisions only on habeas corpus as to the limited question of whether the military had jurisdiction over the court-martialed person.[29]

Its members are now "judges," but the Tax Court, originally the Board of Tax Appeals, is sometimes said to be the only judicial body not in the judicial branch. Its judges are based in Washington, D.C., although masters or judges hold hearings at other locations. The Tax Court shares jurisdiction over tax litigation with the Court of Claims and the federal district courts. The taxpayer who has overpaid must go to the Court of Claims or the district courts. The Tax Court is the only one to which a taxpayer can bring a case before making full payments of taxes instead of paying and suing to recover, but its jurisdiction is limited to cases where the government is trying to obtain a deficiency the taxpayer contests.

The federal regulatory agencies dealing with various areas of the economy* have the power to adjudicate cases arising under the statutes they were created to implement. Intended to provide expertise and to withdraw regulatory cases from judges hostile to regulation, the commissions are composed of from five to eleven individuals appointed for long, staggered terms. The commissions and executive branch agencies with regulatory functions really serve as an adjunct to the national court system because they decide far more cases than the district courts decide. There are at least ten times more cases determined after a formal hearing before an agency than there are civil trials in all federal district courts, and this does not

*Included are all the so-called "alphabet soup" agencies—Interstate Commerce Commission (ICC), Federal Trade Commission (FTC), Federal Power Commission (FPC), Federal Communications Commission (FCC), National Labor Relations Board (NLRB), Securities Exchange Commission (SEC), Civil Aeronautics Board (CAB), and Nuclear Regulatory Agency (NRC), formerly the Atomic Energy Commission (AEC). Many executive branch agencies—including the Departments of Agriculture; Health, Education, and Welfare; Labor; and Transportation —also perform similar functions.

include the many administrative agency actions which occur without hearings—perhaps over 90 percent of the agencies' work. Formal hearings are conducted by administrative law judges, formerly called hearing examiners. The procedures they use have become more and more courtlike, and they rely heavily on the Federal Rules of Evidence. Appeals from the administrative law judges' decisions go to the commissioners, who tend to prefer regulating through a case-by-case approach instead of using the board rule-making (quasi-legislative) powers granted them by Congress. Because review of agency action is limited in the courts of appeals where most appeals from agency rulings go, the agencies are in effect the trial courts for matters under their jurisdiction. Although the agencies' position is well established, some lawyers continue to feel that courts, not "administrative" bodies, should apply the law. From time to time, they suggest that an administrative court be created at the court of appeals level to receive all agency appeals.

JUDICIAL ADMINISTRATION

Judicial administration is a low-visibility activity, yet neither any single court nor the judicial system as a whole could function without it. The federal courts' increasing caseload and Chief Justice Burger's efforts, in addition to focusing more attention to the subject, have helped produce a desire to shift from haphazard and inefficient management to more organized and "professional" administration. For many years, little attention was given to the personnel necessary to make the courts operate effectively. Hiring of administrative personnel was haphazard and essentially a matter of the judges' patronage. This was certainly true of the clerks of court, who until recently were fully responsible for administering court activities. However, the separate position of court administrator is now being created, and each court of appeals has a circuit executive. The people hired for these positions are often professionally trained, particularly at the Institute for Court Management, set up at Chief Justice Burger's suggestion with private funding. The Chief Justice has also appointed an administrative assistant to assist him with his administrative tasks for the Court and the judicial system as a whole.

Before greater attention was paid to court personnel, some concern was shown about the law clerks, who serve not as administrators but as research assistants and opinion drafters for the judges. Usually recent top-ranked law school graduates, they serve for a year or at most two. However, the Supreme Court, where each justice is entitled to three clerks, has hired some career law clerks to provide expertise and continuity, and the courts of appeals are hiring staff attorneys who work for the court as a whole. Despite their importance to the performance of individual judges' tasks, law clerks did not appear on a salaried basis until the mid-1940s in the district courts. And court reporters, who record and transcribe what is said in court, were not hired by the courts on a uniform basis until 1944; before then, most courts left it to the litigants to arrange for private reporters.[30]

To define "judicial administration" in terms of supervision of personnel is too narrow. The directing of activities related to case processing must also be included.[31] From this perspective, the basic units involved in administering the federal judicial system are the Judicial Conference of the United States, the Administrative Office of the United States Courts, and the circuit conferences and councils. The Chief Justice also plays an important role and has done so from the early years when he "received information on the state of judicial business in the far-flung districts, commented on the quality of jury charges given by district judges, and interpreted recent Supreme Court decisions for the benefit of uncertain lower court magistrates."[32]

The system's basic units were established relatively late in the history of the national judicial system—a result largely of the system's "administrative politics" and "judicial politics." Federal judges have been able to remain largely independent of each other and of any central authority. Thus even things as apparently simple as uniform personnel and salary systems for the courts were difficult to establish. Every effort to assert central control has been met with resistance, including lobbying by judges for their own policies and other attempts to seek assistance from members of Congress. The difficulty in developing rational national judicial administration is shown by the problems in developing methods for assignment of judges to handle cases where their help is needed. Until 1850, a federal judge could not sit outside his district. After 1850, he could be assigned only to a court within his own or a contiguous circuit—and only to help a sick or disabled judge. In the early twentieth century, the Chief Justice could ask that a judge serve elsewhere, but the chief judge of the judge's circuit could refuse the request —and did so. Judges, as a result of their primarily local or state orientation, did not see themselves as part of a national judicial system.

Problems surrounding intercircuit assignment of judges have changed— it is not as easy to block requested transfer—but have not disappeared. If all courts have crowded dockets, it is difficult to find anyone except senior (semi-retired) judges to assist with particularly overloaded districts. Yet a "pool" of federal judges available for assignment anywhere has never been established, although the idea was proposed by Chief Justice Taft. Chief Justice Burger has again raised the idea, both because of Congress's failure to provide more judgeships and because of two- and three-year delays in filling existing vacancies. However, given the local basis of judges' appointments, it seems unlikely to be adopted soon.

Judicial Conference

The establishment by Congress in 1922 of the Judicial Conference—originally the Conference of Senior Circuit Judges—was the first official manifestation that authority was being centralized. The Chief Justice chairs this twenty-four-member group, which consists of the chief judge of each court of appeals, a district judge from each circuit (not part of the original membership), and the Chief Judge of the Court of Claims. The Conference was created to be the chief administrative policy maker for the federal judiciary but "constituted only a first step toward a more integrated administrative system."[33] At first it operated mainly by tying the senior circuit judges into a national communications network. It never developed the predominance for which it was intended, in part because its infrequent meetings reduced its potential strength, as did the later development of committees, which proliferated and became more important under Chief Justice Stone. (By the mid-1970s there were almost twenty standing committees plus assorted other groups.) Development of the committees left the Conference largely as a ratifying body for committee-recommended policies. Such policies covered a wide variety of matters, such as rules of evidence and procedure for the federal courts—discussed earlier in connection with the Supreme Court's rulemaking power—the intercircuit transfer of judges just noted, and the need for additional judges to deal with the ever-growing caseload of the federal courts. Although the growth of committees did serve to decrease the power of the Conference itself, its authority was still substantial, and it dealt with matters like ethical standards for judges and other court personnel, qualifications for those personnel, bankruptcy administration, and, most important, the budget for the federal judiciary.

The judges representing the Conference and its committees and the Director of the Administrative Office present legislative proposals to Congress. This allows the Chief Justice to stay away from a direct role in that area, which can cause trouble. When Director of the Administrative Office Rowland Kirks, in the company of an interest group lobbyist, visited Speaker of the House Carl Albert in late 1972 about a bill concerning product safety, Chief Justice Burger was accused of sending a lobbyist to argue against the bill. Denying the charge, the Chief Justice said his only concern was the bill's possible effect on the court system—part of his desire for "court impact statements."

On the whole, the authority of the Chief Justice has been substantial. At first his authority was enhanced because he had to speak for the Conference between meetings. Later it was further strengthened because only he and not the other justices participated in the Conference.[34] The present administrative role of the Supreme Court as a whole is further reduced by the provision that the director of the Administrative Office would perform duties assigned by the Supreme Court and the Conference acting together, although the Chief Justice has continued to be important after the director is named. This is true even when a Chief Justice like Earl Warren gives the director a relatively free hand, and is particularly true where there is a close working relationship, as there has been between Chief Justice Burger and the director.

Circuit Councils

The basic administrative unit of the federal judicial system is supposed to be the judicial *council* in each circuit, composed of the circuit's active judges. (Each circuit also has a judicial *conference*, which includes district judges and lawyer representatives, which meets once a year and may consider a wide range of matters.) Among the councils' wide variety of powers are approving judicial accommodations, directing where court records should be kept, and performing other "housekeeping" tasks; consenting to the assignment of judges to courts in other circuits and dividing judicial business in the districts if the district judges cannot agree among themselves; and the far more sensitive business of certifying to the president that because of disability a judge is unable to discharge his duties. The council also receives reports from the Administrative Office on the dockets in the circuit's courts and is supposed to take appropriate action on the basis of those reports.

One reason for establishing the circuit councils was to decrease the chance of an attack on the central judicial establishment in Washington. Yet diffusing responsibility among all the judges of the circuit has not produced a willingness to act on important matters or to exercise authority, perhaps because the councils "are required ... to consider the behavior of judges with whom they interact on a judicial and sometimes even personal basis." For this reason and because the councils are more attached "to the ideal of local self-government and an independent judiciary,"[35] the circuit councils have played a less important role than the national Judicial Conference.

Problems of circuit council–judge relations and the Supreme Court's hesitancy to deal with them became quite visible in "the liveliest, most controversial contest involving a federal judge in modern United States history,"[36] between the Tenth Circuit's Judicial Council and District Judge Stephen Chandler of the Western District of Oklahoma. The council had removed Chandler—a defendant in civil and criminal lawsuits and the object of attempts to disqualify him from sitting in several cases—from deciding any pending or future cases. Chandler, not wanting to acknowledge the council's authority, did not appear at a council hearing on the matter. The council then left Chandler his pending cases, but distributed new cases to other judges in the district. Chandler's objection to the council's action was that because the Constitution provided nothing short of impeachment, impeachment was the only way to proceed against a judge. (He obviously disagreed with the increasing although still small number of people who feel that a federal judge's appointment "during good behavior" allows suspension or removal of a judge without the impeachment process, reserved for "high crimes and misdemeanors.")

When Chandler tried to obtain a writ of prohibition and mandamus against the council from the Supreme Court, the Court used jurisdictional grounds—his failure to seek relief from either the council or some other court—to rule against him, and thus did not deal with the substance of his claim. Chief Justice Burger, speaking for the Court, did, however, get in some strong language on the need for administration of court dockets:

> There can ... be no disagreement among us as to the imperative need for total and absolute independence of judges in deciding cases or in any phase of the decisional function. But it is quite another matter to say that each judge in a complex system shall be the absolute ruler of his manner of conducting judicial affairs.

Justice Harlan, concurring in the Court's action, felt the Court should face the issues Chandler presented and said the council's orders were within its authority. Justices Douglas and Black, however, dissented strongly. Feeling that the Court was being unfair to Chandler in forcing him to recognize the council's authority if he wished relief, they agreed with him that impeachment was the only way to remove a judge, and argued against the circuit council control used against Chandler:

> It is time that an end be put to these efforts of federal judges to ride herd on other federal judges. This is a form of "hazing" having no place under the Constitution.... If they break a law, they can be prosecuted. If they become corrupt or sit in cases in which they have a personal or family stake, they can be impeached by Congress. But I search the Constitution in vain for any power of surveillance which other federal judges have over those aberrations.

The Department of Justice and the Administrative Office

Since its creation in 1870, the Department of Justice has performed day-to-day administrative tasks for the federal courts. The department's involvement was not ended by the formation of the Judicial Conference. Pervasive conflict over administration has taken place between the department and individual judges. Department field auditors visiting judicial districts to look over books, records, and accounts performed "a communication function as well as an investigative one."[37] However, clerks of court resisted department demands for the reporting of financial transactions. Despite such opposition, the department increased its control over district court clerks and their deputies in 1919, courts of appeals clerks in 1922, and the probation service in 1925. Yet obtaining statutory authority over particular functions did not lessen the need for the department to fight to obtain actual control, and its failure to submit recommendations of the Conference of Senior Circuit Judges to Congress or to support them when submitted did not win the judges' favor.

The department still performs an important role for the judiciary, although not so much in an administrative capacity. The ranking officials of the department are very important for the judicial system. The solicitor general argues the government's cases before the Supreme Court, but more important are his decisions as to whether to appeal a case the government has lost in the lower courts. At the district court level, the United States Attorneys exercise considerable discretion in determining what cases they will bring. They are under the formal supervision of the attorney general, who can control politically sensitive cases or those in certain categories such as civil rights or can instruct the attorneys as to whether or not to try a case. In most cases, however, formal supervision is not matched by actual control of the prosecutors' work. That work is affected more by the prosecutors' interaction with district judges and with the other federal agencies which initiate cases and on which they may need to rely for information.[38] The U.S. Attorneys' offices vary widely in size, from the U.S. Attorney and one assistant to an office (the Southern District of New York) with over one hundred assistants. Although formally nominated by the president and confirmed by the Senate, the U.S. Attorneys are in reality chosen by the senators and thus have a local orientation. Chosen similarly are the U.S. marshals. They and their deputies serve as administrative officers for the district courts. They, too, are administered under the attorney general's control.

Despite the department's importance, considerable change in the administrative "center of gravity" occurred in 1939, when the Administrative Office of the Court was established by Congress in the aftermath of the Roosevelt "Court-packing" plan, in which FDR, proposing to reform judicial functioning, had suggested that a proctor (court administrator) be appointed for the Supreme Court. When the Administrative Office (A.O.) was first established, members of the Judicial Conference acted to limit the authority of the office and its director, appointed by the Supreme Court. It was the Conference's intent that the A.O. be "an executive office with strictly limited power," dealing with judges and other court personnel only upon authorization of the Conference, circuit councils, or individual courts, and performing housekeeping functions for the courts without employing the courts' personnel. However, the A.O. has often initiated policies later considered by the Conference and its recommendations have carried great weight.[39]

The Justice Department still performs field audits of court accounts, but the Administrative Office took over most other department-performed functions plus some which had not been performed at all by the department.

The A.O. has the duty to compile suggested budgets and prepare vital statistics, examine dockets, determine personnel needs and procure supplies, and carry out additional tasks the Supreme Court or the Judicial Conference assigns, for example, establishing federal public defender systems to implement the Criminal Justice Act. The A.O. also serves as the secretariat for the Judicial Conference and its committees and as the liaison between the judiciary and Congress, individual judges, professional organizations, and other government agencies.

The Federal Judicial Center is the most recently created (1967) element in federal judicial administration. It was set up to meet research and educational needs of the judiciary. Among the tasks it has undertaken have been testing calendaring procedures, studying case screening and delays in the filing of transcripts, preparing a manual for complex and multidistrict litigation, and conducting training programs for magistrates, newly appointed district judges, and probation officers.

In this chapter we have examined the basic elements of the federal judicial system, looking at some basic structural aspects of the Supreme Court which are to be examined more fully in later chapters and then examining the other courts in the system—the courts of appeals, district courts, magistrates, and specialized courts, as well as taking a brief look at the federal regulatory agencies. The often neglected subject of the administration of that system was then examined, with attention to the structures by which judicial administration is carried out, the politics of the establishment and development of those structures, and the working relationships between the units. With this picture of organization of the courts established, we now turn to an examination of the process by which the judges of the lower courts and the justices of the Supreme Court are selected and their backgrounds and qualifications.

NOTES

1. *North Georgia Finishing Co. v. Di-Chem*, 419 U.S. 601 at 615–19 (1975). The earlier two cases are *Fuentes v. Shevin*, 407 U.S. 67 (1972), the 4–3 ruling, and *Mitchell v. W. T. Grant Co.*, 416 U.S. 600 (1974).
2. The case was *Laird v. Tatum*, 408 U.S. 1 (1972); Rehnquist's statement is at 409 U.S. 824 (1972). See John MacKenzie, *The Appearance of Justice* (New York: Scribner's, 1974).
3. See Walter Murphy, *Congress and the Court* (Chicago: University of Chicago Press, 1962), for the complete story.
4. *Ohio v. Wyandotte Chemical*, 401 U.S. 493 (1971); *Illinois v. City of Milwaukee*, 405 U.S. 91 (1972); *Washington v. General Motors*, 406 U.S. 109 (1972).

5. *Arizona v. New Mexico,* 96 S.Ct. 1845 (1976).
6. In a recent interesting exercise of its original jurisdiction, the Court had to decide whether it should accept a boundary agreement between contesting states without making an independent assessment of the legality of elements of the agreement, but the majority held, over three dissents, that it could. *New Hampshire v. Maine,* 96 S.Ct. 2113 (1976).
7. Howard Lesnick, "The Federal Rule-Making Process: A Time for Reexamination," *American Bar Association Journal* 61 (May 1975):582.
8. See William L. Hungate, "Changes in the Federal Rules of Criminal Procedure," *American Bar Association Journal* 61 (October 1975):1203–207.
9. *Whalen v. Roe,* 96 S.Ct. 164 at 166 (1975) (Justice Marshall).
10. Kenneth N. Vines, "The Role of the Circuit Courts of Appeal in the Federal Judicial Process: A Case Study," *Midwest Journal of Political Science* 7 (November 1963): 309.
11. Richard J. Richardson and Kenneth N. Vines, *The Politics of Federal Courts* (Boston: Little, Brown, 1970), pp. 118–19.
12. Vines, "The Role of the Circuit Courts of Appeal," p. 314. See also Richard J. Richardson and Kenneth N. Vines, "Review, Dissent and the Appellate Process: A Political Interpretation," *Journal of Politics* 29 (August 1967): 606.
13. J. Woodford Howard, "Litigation Flow in Three Circuits," *Law and Society Review* 8 (Fall 1973): 46.
14. See Russell R. Wheeler, "The Supreme Court's Workload and the Demands on the Federal Appellate Courts," paper presented to the Midwest Political Science Association, 1975, p. 2 of footnotes; derived from data from Administrative Office of the Courts.
15. Howard, "Litigation Flow," pp. 46, 44.
16. For a discussion of these points, see Paul D. Carrington, "Crowded Dockets and the Courts of Appeals: The Threat to the Function of Review and the National Law," *Harvard Law Review* 82 (January 1969):581–84, and Richardson and Vines, *The Politics of Federal Courts,* p. 125.
17. Howard, "Litigation Flow," p. 40.
18. For the report of the Freund Study Group, see *American Bar Association Journal* 59 (February 1973):139–44; Freund's defense of the proposal is in *American Bar Association Journal* 59 (March 1973): 247–52. For criticism, see, inter alia, Eugene Gressman, "The National Court of Appeals: A Dissent," *American Bar Association Journal* 59 (March 1973): 253–60.
19. If the District of Columbia Court of Appeals has heard a case involving a law "not applicable exclusively to the District of Columbia," the case still goes to the U.S. Court of Appeals.
20. *Palmore v. United States,* 411 U.S. 389 (1973).
21. *Gonzalez v. Automatic Employees Credit Union,* 419 U.S. 90 at 98 (1974). See also *MTM, Inc. v. Baxley,* 420 U.S. 799 (1975).
22. Data from the Administrative Office of the Courts, Annual Report of the Director. For dispositions of district court caseload, see also Sheldon Goldman and Thomas Jahnige, *The Federal Courts as a Political System,* 2nd ed. (New York: Harper and Row, 1976), pp. 116–21 (civil), 121–29 (criminal).

23. Gregory J. Rathjen, "Population Growth and the Federal Judicial System," *Political Issues in United States Population Policy*, edited by Virginia Gray and Elihu Bergman (Lexington, Mass.: Lexington Books, 1974), pp. 101–24. How federal district court litigation rates in general do not necessarily parallel increasing economic or population growth is discussed in Joel Grossman and Austin Sarat, "Litigation in the Federal Courts: A Comparative Perspective," *Law & Society Review* 9 (1973):321–323.

24. *Thermtron Products v. Hermansdorfer*, 96 S.Ct. 584 (1976).

25. Except for the discussion of cases, this section of magistrates draws extensively on Steven Puro, "United States Magistrates: A New Federal Judicial Officer," *Justice System Journal* 2 (Winter 1976), 141–156.

26. Challenges to the changes were denied in *Glidden v. Zdanok* and *Lurk v. United States*, 370 U.S. 530 (1962).

27. See Symposium, "The United States Court of Claims," *Georgetown Law Journal* 55 (December 1966):393–553.

28. See Lawrence Baum, "Decision-Making in a Specialized Court: The Court of Customs and Patent Appeals," paper presented to the Midwest Political Science Association, 1975.

29. See *Burns v. Wilson*, 346 U.S. 137 (1953).

30. Peter Graham Fish, *The Politics of Federal Judicial Administration* (Princeton, N.J.: Princeton University Press 1973), p. 230.

31. See Russell Wheeler and Howard R. Whitcomb, "What Is Judicial Administration?" in *Perspectives on Judicial Administration: Readings in Court Management and the Administration of Justice* (Englewood Cliffs, N.J.: Prentice-Hall, 1976).

32. Fish, *The Politics of Federal Judicial Administration*, p. 8. This section draws on Fish's definitive presentation.

33. Ibid., p. 39.

34. Ibid., p. 50. For a discussion of the Chief Justice's role in improving administration, see William F. Swindler, "The Chief Justice and Law Reform, 1921–1971," *The Supreme Court Review* 1971, edited by Phillip Kurland (Chicago: University of Chicago Press, 1971), pp. 241–64.

35. Fish, *The Politics of Federal Judicial Administration*, pp. 406, 308.

36. *Chandler v. Judicial Council of the Tenth Circuit*, 398 U.S. 74 at 130 (1970). For a thorough treatment of the entire Chandler matter, particularly the underlying conflict between Judge Chandler and his fellow judges, see Joseph C. Goulden, *The Benchwarmers: The Private World of the Powerful Federal Judges* (New York: Ballantine Books, 1974), pp. 234–84.

37. Fish, *The Politics of Federal Judicial Administration*, p. 93.

38. On this point, see Robert L. Rabin, "Agency Criminal Referrals in the Federal System: An Empirical Study of Prosecutorial Discretion," *Stanford Law Review* 24 (June 1972):1036–91. On selection of United States attorneys, see James Eisenstein, *Politics and the Legal Process* (New York: Harper and Row, 1973), pp. 36–42, 47–49, and pp. 157–70 on the operation of their offices.

39. Fish *The Politics of Federal Judicial Administration*, p. 191.

4 | THE SELECTION OF JUDGES

SELECTION: THE FORMAL PROCESS

Federal judges—nominated by the president and confirmed by the Senate —are chosen by a political process involving both formal procedures and relatively well-defined patterns of participation by nongovernmental groups. While most nominees for federal judgeships are confirmed, the process, particularly prior to nomination, can be far from routine and rejections of nominations, even to the Supreme Court, indicate the importance of the process's nonformal aspects.

The Constitutional Convention debated several methods of selecting the justices. These included selection by the Senate alone, by the entire Congress, and by the executive alone. The method ultimately adopted was the one already in use in Massachusetts—nomination by the chief executive with the advice and consent of the Senate. The Constitution specifies no formal process by which the president is to choose those he will nominate. However, the Department of Justice has come to perform the functions of identifying and "screening" candidates for the president, as people seek judgeships for themselves or for friends and some seek to keep others from the bench. The FBI conducts an intensive investigation on those being seriously considered, with Department of Justice officials evaluating the FBI report. (Prior to the Nixon administration, a background check was run only on the person already selected, but this provided insufficient information.) Once the president's nomination is sent to the Senate, it is referred to the Committee on the Judiciary. After holding hearings to take testimony from the nominee, supporters, and opponents, the committee makes a recommendation to the full Senate, which then debates and votes.

Usually a nominee to a federal judgeship does not assume office until after the Senate has confirmed him. However, the president may make a "recess appointment" to a vacant judgeship, allowing the person to take office immediately. Such appointments are valid, that is, the judge's salary may be paid, only under certain circumstances: if the vacancy occurred within the last thirty days of the Senate session or when the Senate was not in session; if the Senate rejected a nomination during the last thirty days of the session and the recess appointee is someone else; or if a nomination pending at the conclusion of a Senate session is not that of a previous recess appointee. Nominations of recess appointees must also be submitted to the Senate within forty days after the Senate's next session begins.

The American method of selecting federal judges is only one of a number of possibilities. Judges of the Japanese High Court are appointed under a system of quotas under which the court's judges must include five career judges, five attorneys, and five men "of learning and experience" (law school professors or lawyer-bureaucrats). While in the United States judges are chosen from among a general pool of lawyers, in many other countries one is either a lawyer *or* a judge; judges have a separate "career track," including separate training. Training of judges in the United States, while increasing, is limited and occurs after selection.

Judges are elected rather than appointed in many of the American states, some using a partisan ballot, some a nonpartisan ballot. Many formally elective systems become appointive in fact where the governor is allowed to fill vacancies, and there is a norm of not opposing "sitting judges." The alternative method of state judicial selection most highly touted by lawyers, who sometimes propose it for the federal courts, is "merit selection" or the "Missouri plan." A judicial selection commission of lawyers and nonlawyers prepares a list of three names from which the governor must make the appointment, with the public voting on whether to retain the judge after at least a year of service. This noncontested election is repeated for each new term.

Attracting, Retaining, and Removing Judges

In the Supreme Court's early years, presidents at times had difficulty obtaining men to serve as justices, because of the job's lack of prestige. Some who did serve left the Court after brief service or simultaneously held other government jobs. Now, however, the position of justice of the Supreme Court is thought to be the highest to be attained in the legal profession and is likely to be accepted readily when offered—although apparently President Johnson's initial offer of a position was refused by Abe Fortas. Justices now only rarely leave the Court for other positions and then, as in the case of James Byrnes to become the president's special assistant during World War II and Arthur Goldberg to be ambassador to the United Nations, only at the president's urging.[1] (Justice Tom Clark retired so that his son, Ramsey Clark, could become attorney general without creating conflicts of interest for both, but the elder Clark continued to serve as a federal judge.)

Attracting lower court judges is a greater problem, in large measure because of salary. Until 1977, Supreme Court justices received $63,000 and the Chief Justice $65,600 (now raised to $72,000 and $75,000), but district judges received only $42,000 and court of appeals judges $44,600; moreover, they had received no raises for more than six years until 1975. Such salaries might seem more than adequate, but many judges could earn substantially larger incomes—well over $100,000—in private practice, and many considered the compensation inadequate, particularly if they had children to put through college. The judges' positions remain prestigious, and some observers argued that there was an ample pool of qualified people who wished to be federal judges even at those salaries. However, several district judges resigned because of the salary situation, and the problem was graphically illustrated by an *American Bar Association Journal* advertisement: "Judge, United States District Court, under 50, seeks increased earnings opportunity in law administration or business."[2]

In 1976, forty-four federal judges (later joined by thirty-seven others) filed suit in the Court of Claims on the basis of the constitutional provision that judges' pay shall not be reduced during their term in office. The judges said their pay had been so eroded by inflation since 1969 that, even with the 5 percent increase, a district judge's salary had been reduced to the equivalent of $27,510 and a court of appeals judge's salary to $29,230. Congress had given the president the power to raise judicial salaries every fourth year and in 1974 President Nixon had proposed a 7.5 percent increase for each of three years, but the Senate had canceled the proposed raises, they noted. The judges of the Court of Claims, before hearing the case, *Atkins v. United States,* asked the Supreme Court (under a procedure called "certification") whether they could decide it because, with all judges affected by a ruling, no judge could decide the case without being involved in a conflict of interest. With three justices disagreeing, the Supreme Court declined to deal with that issue and the Court of Claims proceeded to hear the suit. In February 1977, substantial pay increases for the federal judges—to $54,500 for district judges and $57,500 for circuit judges—went into effect when Congress did not reject recommendations submitted by President Ford. At least some of the judges continued to claim that the new salary did not meet the effect of inflation and they did not drop their suit. The Court of Claims ruled, however, that the Constitution provided no protection against such "indirect" reduction in judges' salaries. Although some observers have thought the suit unseemly, it did help focus attention on the pay problem, and many believe that was the principal reason for the suit's initiation.

Obtaining and retaining judges is a current problem; getting others—old, ill, or otherwise not functioning effectively—to retire was an earlier one. Indeed, many Supreme Court justices have died in office. While some states have a mandatory retirement age and an increasing number have judicial discipline commissions which can recommend removal, the U.S. Constitution has no such provisions. Although some federal judges have been impeached, convicted, and removed from office, impeachment has been thought too difficult to accomplish or inappropriate in instances of illness or senility, and persuasion by a judge's colleagues has also not been effective. (Only one Supreme Court justice, ardent Federalist Samuel Chase, was impeached, and he was not convicted.[3]) Attention thus turned to development of incentives to retirement through an effective retirement system.

Prior to 1937, a judge could retire on his then not-very-generous salary after ten years of service and on reaching age seventy. Congress changed the plan in 1937 so that a judge at that age and amount of service could opt for partial retirement. (In 1954, the law was amended to provide the option at fifteen years service and age sixty-five.) This senior judge status makes the judge, who continues to benefit from any pay raises received by those in full-time service, available for duty while the president names a new judge to the judge's original position. Because of the lower federal courts' substantial caseloads, senior judges—some of whom work almost as hard as before they "retired" and who make themselves available for service in other districts or circuits—are necessary to the work of the federal judicial system.

THE LOWER COURTS

Selection and Confirmation

In actuality, under the practice of "senatorial courtesy," the president shares the selection of district court judges with the senator(s) from the president's party in the state in which the appointment is to be made. (If there are no senators from the president's party, the state party organization may be consulted.) Most senators of the president's party feel they should designate the nominee, and "some even take a proprietary view, that they own the job,"[4] although others may do no more than submit a list of acceptable nominees, leaving the choice to the president.

A senator once could assert senatorial courtesy with little challenge simply by stating that—whatever the real reason—an opposed nominee was "personally obnoxious," but objections must now be substantiated. Moreover, senatorial courtesy does not assure a senator that his own nominee will be accepted when conflict with the president occurs. It does, however, mean that the senator can probably block the president's nominee—evidence that a senator's power is largely defensive. (Senators losing such battles have done so because the successful appointee was sponsored by the state's other senator of the same party.) A president, because he will later need senators' votes for his program, generally does not challenge senatorial opposition. Yet the president's formal constitutional power to make judicial nominations provides him an advantage. So does the fact that some senators of their own choice play only a minor role in the selection process, not asking for a specific nomination because they do not wish to owe the president a favor; similarly, many senators do not wish to trade a vote on an issue in return for approval of a nomination.

For court of appeals appointments, the effect of senatorial courtesy is substantially diminished if not fully eliminated because they do not fall within a single state, although a tradition that certain seats "belong" to particular states strengthens the senators' hands. At a minimum, a senator may demand geographical parity within the circuit. (Senator Hiram Fong, R-Hawaii, a member of the Judiciary Committee, was able to delay several judicial nominations until someone from Hawaii was nominated to the Court of Appeals for the Ninth Circuit.) Although senatorial courtesy does not operate with full force for the appeals courts, strong protests from a senator from the nominee's state may be needed. For example, when President Kennedy planned to promote District Judge J. Skelly Wright, who had ordered desegregation of the New Orleans schools, to the Court of Appeals for the Fifth Circuit, complaints from Louisiana's senators led the president instead to appoint him to the Court of Appeals for the District of Columbia, where, because there are no senators, senatorial courtesy cannot operate. (For Supreme Court nominations, senatorial courtesy is virtually never operative, although the sponsorship or at least acquiescence of the senators from the nominee's state is helpful.)

Once a nomination is made, senators other than those from the nominee's state may become involved. Although not inclined to interfere with district court nominations in other states, senators may have to take an interest in some situations if they are serious about the quality and image of the federal judiciary. Interest groups supporting a senator, particularly if the group does not have favorable access to the executive so that it could block a nomination there, may also bring pressure. The largest role in the process is, of course, played by the senators on the Committee on the Judiciary. A particularly stubborn committee chairman, like Senator James Eastland (D-Miss.), can force a hard bargain—like ending the delay on Thurgood Marshall's appointment to the Second Circuit in return for the district court nomination of a former law partner. The committee has the power to test the president's will by delaying action on nominations. In return for approving administration choices, its members may also be able to gain nominations they wish, particularly when large numbers of judges are to be named at one time, for example, when the number of judgeships is increased. In such situations, prior arrangements over allocation of the positions may be necessary to obtain congressional approval.

Through Judiciary Committee hearings, the entire Senate, the media, and the general public can be attuned to objections against a nominee. The committee can decline to let someone testify, relegating objections to the less visible printed record, and it can control the sequence in which witnesses do appear and the treatment, hostile or friendly, they receive. The committee members also lead debate on the Senate floor, "another opportunity for senators to seek to embarrass the administration by questioning the wisdom of a particular appointment."[5] Anticipating that such questions will be raised, a president may prefer to avoid a particular nomination or even to withdraw it once made despite the resulting embarrassment.

The selection process as it presently operates took form during the Eisenhower administration. Prior to the formal nomination, the Department of Justice and the president's staff perform vital functions in the selection process. However, their role differs from one presidential administration to the next, as does that of the American Bar Association (ABA), whose participation is more significant in Republican than Democratic administrations.[6] President Eisenhower publicly committed himself to quality judicial appointments and took particular responsibility for them. However, others handled the work, with principal responsibility in the hands of the deputy attorney general—in the Eisenhower administration, as in the Ford administration, a former judge.

The Eisenhower administration explicitly stressed previous judicial experience (particularly for appeals court positions), age, and the ABA's ratings as crucial factors in selection. The average age at appointment was, however, essentially the same for the Truman, Eisenhower, and Kennedy administrations, although Eisenhower did lower the maximum age at which a person would be considered, so that no district judge appointee was over sixty-three and no one over sixty-six was named to the courts of appeals. The Kennedy administration said less about the age factor but lowered the age more than had Eisenhower. Similarly, it almost matched the Eisenhower record in choosing people with previous federal judicial experience for the courts of appeals. On this point, Eisenhower did not do as well as the Truman administration had done and did far less well than the Hoover administration, in which eleven of sixteen appeals court appointees came from the district courts. However, promotion of federal district judges to the courts of appeals is likely to be limited when the presidency changes hands because a president is simply not likely to elevate district judges appointed by a president of the opposite party. District judges who move to the courts of

appeals thus are likely to do so within a short time of their initial appointment or not at all—to be moved forward by the administration which first
appointed them to the federal bench. The Eisenhower administration also
did not give state appellate judicial experience more weight than past administrations, although a nominee's experience as a trial judge or trial lawyer
appeared to be of considerable importance.

The ABA has been involved in the judicial selection process at the national level since 1946. It had sought to obtain review of judicial nominations
before they were made public and preferably even before they were definite. (At the committee hearing stage, the most the ABA can do if it has not
blocked a nomination earlier is to lodge an objection with the committee.)
Eisenhower's emphasis on "quality" led to increased ABA involvement,
which later administrations could not easily avoid, although the group's
participation was controversial. (Opponents have objected to the ABA's
special access to those making the appointments and have attacked the
group's conservatism, said to be masked by its claim to be interested only
in judicial competence.) During the Eisenhower administration, the Department of Justice obtained the views of the ABA's Committee on Federal
Judicial Selection before a nomination was decided, and in Eisenhower's last
two years the committee "had a virtual veto power," so that only those the
committee had rated Qualified were nominated.[7] (The committee's ratings
are Exceptionally Well Qualified, Well Qualified, Qualified, Not Qualified,
and Not Qualified by Reason of Age.)

The Kennedy administration did try to limit the ABA committee's role.
Top Justice Department personnel were unwilling to give in to the ABA
when they differed with its committee. Indeed, using "argument and cajo-
lery,"[8] the administration was able to get the ABA to change some of its
ratings. Almost one-third (29 percent) of the ABA's informal ratings differed
from formal ratings finally announced. In nearly one-third (30 percent) of
the changes, the rating was *up*graded, perhaps as a result of the ABA's desire
to appear successful when the administration was intent on a particular
nomination.

The Kennedy administration's approach to judicial nominations was also
generally more casual and informal than that of Eisenhower. Kennedy had
far less interest in judicial appointments than in social programs despite the
judges' potentially great impact on those programs, such as desegregation,
severely hindered by some southern district court appointees.[9] Extended
delay in filling a position was used in nineteen of a hundred appointments
to bring senators around to the president's choice. Recess appointments were
also used more frequently, accounting for one-fifth of the nominations,
twice the proportion in the Eisenhower period, but most were not used to
pressure senators. The Johnson administration showed more deference to
senators' wishes, at first because of Johnson's attention to other matters but

mostly because of Johnson's background as Senate majority leader. Less heed was paid to the ABA at first. However, Johnson received much "bad press" for his nomination to the Fifth Circuit Court of Appeals of Governor James P. Coleman of Mississippi, attacked for his racial views, and for his nomination to a district judgeship of Kennedy crony Francis X. Morrissey, so obviously unqualified that the nomination was ultimately withdrawn. The extremely negative reactions to these choices led Johnson to be more sensitive to ABA reaction, which thus improved. The ABA committee ranked six of Johnson's first fifty-six appointments to the district and appeals courts Not Qualified, but by 1966 the committee thought the administration's appointments quite good.

During the Nixon presidency, the ABA's position again strengthened. The president said he would not name someone considered Not Qualified, but late in his administration he broke his word by nominating Connecticut Governor Thomas Meskill to the Second Circuit Court of Appeals despite ABA opposition—perhaps because the vacancy was a "Connecticut seat" and Meskill's sponsor, Senator Lowell Weicker, appeared to have little regard for the ABA. The ABA's importance in lower court nominations continued into the Ford administration, and the bar group seemed quite pleased—at least publicly—with its relations with the administration.

Where the ABA would stand in the Carter administration's judicial selection scheme was unclear, although if history were to repeat itself the bar group's influence would decrease. However, that President Carter intended to change the selection process was clear from his executive order establishing merit selection commissions (of eleven members each, including lawyers and laypeople) for each circuit to provide him with names of the five best qualified candidates for each court of appeals vacancy. The parallel establishment of similar screening groups in some of the states for district court judgeships—the group for California even included some members named by the state's new Republican senator—made even clearer that the new president was going to try to alter the selection process.

Judicial Backgrounds

Interest in judges' background characteristics stems from the supposed relation of those characteristics to judges' patterns of voting in cases. However, except for political party, which reflects differences in socioeconomic status and ideology, such relationships—which are clearer at the state level where judicial selection and advancement processes are often more openly partisan—appear fuzzy for federal judges. Goldman found some differences between Democrats and Republicans on the federal courts of appeals, for example, in the area of economic policy, but not in other areas. Other factors which at first seemed related to circuit judges' voting "washed out" when political party was taken into account, leaving it the single most helpful explanatory factor.[10] By comparison, Nagel found that U.S. Supreme Court and state supreme court justices with prosecution backgrounds or with ABA membership voted less for the defense in criminal cases and that Protestant, Anglo-Saxon judges were more conservative than others. Party affiliation was more important than ethnic background in predicting votes: Democratic judges chose the more liberal response in fifteen policy areas examined, with the differences between Democrats and Republicans statistically significant in nine of them.[11]

The selection process for the lower federal courts is very likely—nine of every ten times—to produce judges of the president's political party. The nominees often have been judges or prosecutors; during the Johnson administration and Nixon's first term, for example, only one-fifth of the appeals court nominees had neither type of experience, with the picture similar for district court judgeships outside the South.[12] Eisenhower's nominees were particularly likely to be from relatively large private law firms or from government practice; high-ranking Justice Department officials accounted for roughly 10 percent of his nominations. More of Kennedy's choices had been either in elective positions or some form of government administration; the private lawyers he nominated were more likley to have been in small firms or individual (solo) practice. The same differences occurred between the (Democratic) Johnson and (Republican) Nixon administrations; 40 percent of Nixon's district court appointees were partners in large law firms, with only 10 percent from small firm and solo practice, while roughly one-fifth of Johnson's appointees came from the latter sources.

More Eisenhower than Kennedy appointees attended college and law school at Ivy League schools for both their undergraduate and legal educations. Outside the East, however, very few judges attended Ivy League undergraduate schools. The higher socioeconomic status of Republicans was perhaps not reflected in their education because when the Johnson and Nixon judges were attending school, the private and Ivy League schools increasingly had been admitting students on the basis of merit and not background. Differences in religion and race also appear between Republican- and Democratic-appointed judges. Over three-fourths of Eisenhower's appointees were Protestant, more than Kennedy's district court and appeals court appointees (three-fifths and two-thirds, respectively). Kennedy appointed correspondingly more Catholics and Jews. Similarly, proportionately more Nixon than Johnson appointees were Protestant; Johnson appointed more Cathlics and Jews. Eisenhower appointed no blacks to federal judgeships; Kennedy named four to the district court and one to the court of appeals. Johnson, in addition to naming Thurgood Marshall as the first black to sit on the Supreme Court, named five blacks to district judgeships and two to the court of appeals; Nixon named only four black district judges. (One LBJ district court nominee was not confirmed; he was later nominated by President Ford and confirmed.) Although Johnson did appoint two women to be district judges and one to a court of appeals, appointments of women to federal judgeships have been negligible. With President Ford's two district appointments of women, there are now only five women active-duty federal judges.

THE SUPREME COURT

The Justices' Qualifications

The qualifications of Supreme Court justices can be categorized as representational, professional, and doctrinal.[13] The principal representational qualification is political party. As with lower court judges, the great majority (90 percent) of the more than one hundred justices have been of the appointing president's party. There have been more justices whose home states were in the East than in the West, although some effort is made to avoid overrepresentation of any part of the country. Representation of minorities has been apparent, although whether it has been purposeful is disputed. A "Jewish seat" seemed to exist once Louis Brandeis joined the Court (anti-Semitism infused much of the opposition to his nomination). Later occupants of that "seat" were Felix Frankfurter, Arthur Goldberg, and Abe Fortas. The difficulty with asserting the existence of a "Jewish seat" is that Benjamin Cardozo, also a Jew, served simultaneously with Brandeis, and Fortas was not replaced by another Jew.

Catholics have regularly served on the Court, but there has been less conscious attention to a "Catholic seat" than to a Jewish one. For example, Eisenhower's nomination of William Brennan resulted primarily from the president's need to find a highly qualified "sitting judge" to quiet criticism of the "political" appointment of Earl Warren, a former Republican vice-presidential candidate, and from the unavailability because of age of Arthur Vanderbilt, chief justice of the New Jersey Supreme Court of which Brennan was a member. No black reached the Supreme Court until Lyndon Johnson's appointment of Thurgood Marshall and no woman has done so, despite Betty Ford's lobbying at the time of Justice Douglas's resignation.

Most of the justices have come from upper-middle-class or upper-class surroundings; very few have been of "essentially humble origin." Into the nineteenth century, justices came from "socially prestigeful and politically influential" families of the gentry class; later, professional family backgrounds predominated and economic rather than political prominence was likely.[14] The nation's more open political system resulting from Jacksonian democracy was not directly reflected in the Court's membership. Members of justices' families were often active in public life, and more than one member of the same family has served on the Court, for example, Stephen Field and his nephew David Brewer (some said this gave Field two votes) and the two John Marshall Harlans, one at the end of the nineteenth century and his grandson in the 1950s and 1960s.

Justices' ages increased as the nineteenth century progressed; this was a result of our political development and the concomitant growth of informal career lines. The likelihood that justices would be older was reinforced by formal law training and greater specialization in the legal profession, so that it took longer to become established in the profession and receive appropriate recognition. While Justice Douglas was forty and Justice Rehnquist in his forties when appointed, appointment to the Court in an individual's late fifties, as with John Paul Stevens, or sixties, true of three of the Nixon nominees, is more typical.

Professional qualifications mean prior judicial and governmental service. The state courts provided most of the Supreme Court justices in the last century, although at the end of the century justices were more likely to have had national rather than local experience. A nominee's judicial experience has always been taken into account, although lack of it has not been a bar to appointment, as in the case of Hugo Black, who had served briefly as a police magistrate but whose primary activity had been as a U.S. senator. The American Bar Association focused more attention on this factor and criticized nominees who lacked judicial service, such as Earl Warren—criticism compounded by his being chosen to be Chief Justice. Warren's experience as a prosecutor and state attorney general as well as governor did, however, serve to diminish the criticism.

Most other recent appointees either had judicial experience, had held important government jobs, or were distinguished lawyers. Eisenhower appointees John Marshall Harlan, Potter Stewart, and Charles Evans Whittaker, Johnson appointee Thurgood Marshall (solicitor general at the time of his appointment), Nixon appointees Burger and Harry Blackmun, and Ford appointee Stevens—a former Supreme Court law clerk—were federal appeals court judges, as were unsuccessful Nixon nominees Clement Haynsworth and G. Harrold Carswell, earlier a district judge. William Rehnquist, also a former Supreme Court law clerk, was an assistant attorney general and Kennedy appointee Byron White was deputy attorney general when appointed. Both Arthur Goldberg and Abe Fortas were distinguished attorneys, as was Lewis Powell, a former president of the ABA.

The effect of these characteristics on the justices' performance is unclear, as it is with lower court judges. While most justices are of the president's party, explicit partisanship is played down and could act as a disqualification. Party affiliation was a much better predictor of voting in the nineteenth century than it is now, because then the parties were more clearly divided on some issues, particularly regional/sectional ones like slavery.[15] A recent study has indicated that in cases involving presidential power where the decisions went one way or the other, the hypothesis that "a greater proportion of the judges from the president's party are likely to decide for him than are the judges from the opposition party" was confirmed in 64 percent of the cases. The justices also tend to vote consistently in support of the policies of the presidents who selected them, with a 68 percent confirmation for the hypothesis that "a greater proportion of judges appointed by the president are likely to vote for him than are the judges not appointed by the president."[16] Such results stem, however, from ideological affinity or "doctrinal qualification" rather than from party affiliation or identification as such.

Turning to past judicial performance, we find that, counter to the bar's expectations, it does not produce a greater reliance on precedent. Instead, justices who had significant pre-Supreme Court judicial experience or who came from families containing judges were more likely to depart from precedent than were justices without such experience or without judges among their relatives.[17] Perhaps those familiar with the judicial role could understand that a judge has to depart from precedent in order to resolve disputes, while those without such experience would be more likely to follow the law school-taught norm of following precedent. Those with prior judicial experience and those from "judicial families" have also been less likely to dissent, maybe because they have accepted the norm of minimal dissent, quite strong in the lower federal courts and state courts. Judges from the more liberal party, who were more likely to have prior judicial experience, were more prone to abandon precedent, but the liberals were also more likely to dissent more frequently than conservative judges. Those most likely to dissent were most often those adhering to an earlier status quo and unwilling to abandon it.

Prior judicial service can help the president identify the nominee's ideological affinities, in which he is most interested. Presidents have been "much more likely to be successful in predicting the future voting behavior of a nominee when the nominee had a record of previous judicial experience that would provide a dependable clue to their probable future behavior on the Court."[18] Thus through professional qualifications the president attempts to ascertain doctrinal qualifications. Presidents have been quite successful in doing this, particularly when we consider their lack of personal knowledge of those they nominate, especially if the nominees are without (extensive) judicial records. "About three-fourths of those justices for whom an evaluation could be made conformed to the expectations of the presidents who appointed them."[19] Although some presidents appear to have few expectations of their nominees, as in the case of Eisenhower with Warren or Brennan, others expend much energy to ensure that someone has been nominated who will vote the "right" way on a key question. Yet even in such situations, the president's expectations are often disappointed. The best-known story involves President Theodore Roosevelt, who selected Oliver Wendell Holmes for the Supreme Court after assuring himself that Holmes would vote "correctly" on antitrust matters. Yet Holmes, in his first antitrust case, voted the "wrong" way, provoking Roosevelt to say he could make a judge with a stronger backbone out of a banana.

A good example of prior judicial service facilitating a president's choice is Chief Justice Warren Burger, like most other Chief Justices selected from outside the Court. Only three nominees—Edward White, Harlan Fiske Stone (neither from the president's party), and Abe Fortas (not confirmed) —have come from within the Court. Charles Evans Hughes had earlier been a member of the Court but was not a member when named Chief Justice. In the policy area in which Nixon was most interested, criminal procedure, Burger's decisions as a court of appeals judge showed a "judicial philosophy ... generally described in terms of strict construction, conservatism, and judicial restraint."[20] Burger had decided against the criminal appellant in twenty-five of twenty-six nonunanimous arrest and search and seizure cases during the 1956–1969 period, and his opinions on the question of "probable cause" to arrest or search reflected "a pro-prosecution orientation which reject[ed] legal technicalities in criminal procedures that restrict the police in conducting their business."[21] Burger's early years on the appeals court (1956–1962) showed a "moderate" record on criminal procedure (he cast 45 percent of his votes for those appealing criminal convictions on procedural issues). After 1962, however, his record became far more conservative at least in relation to his colleagues. Thus in the fourteen *en banc* criminal procedure rulings of the District of Columbia Circuit between 1965 and 1969, he did not vote once in favor of the appellant.[22]

Burger's stance and that of President Nixon's other nominees was quite close to the president's position on criminal procedure. They also often showed the judicial self-restraint and belief in government officials' good intentions that Nixon wanted. Burger's statement that the Court did not exist to "cure every disadvantage human beings can experience"[23] indicates their general approach. Despite this, the Nixon nominees failed to reflect the president's positions on abortion, aid to parochial schools, important aspects of desegregation, and electronic surveillance and in the end handed down the ruling—in the "tapes case"—which led to his departure from office.

Prior judicial service not only helps the president select nominees but also, once a nomination is made, provides others an opportunity to find disliked decisions. Most recently, the National Organization for Women (NOW) objected to the nomination of John Paul Stevens because of some of Stevens's opinions as a judge of the Court of Appeals for the Seventh Circuit. The judicial actions of John Parker and Haynsworth and Carswell, the Supreme Court nominees rejected in the twentieth century, were held against them by labor and civil rights groups.[24] Parker and Haynsworth were said to be too conservative, particularly on policy toward labor unions and on desegregation. (The latter charge concerning Parker was, however, based on a campaign statement.) Defenders of Judge Parker said his judicial decisions were merely implementing the Supreme Court's rulings, which as a lower federal judge he was expected to follow. Although the Fifth Circuit had been more liberal than Haynsworth's Fourth Circuit, his rulings were said not to have evaded or obstructed school desegregation. It was also argued in his behalf that, even though the Supreme Court later overruled him when it began to demand more action, his decisions were an accurate reflection of the Supreme Court's position at the time he made them.

Carswell's service as a district judge produced complaints that he had handled civil rights cases with hostility and had been rude to black lawyers and black defendants. Because the Court of Appeals for the Fifth Circuit had reversed Carswell more frequently than all but a few other district judges in the circuit, his professional competence was questioned; his supporters asserted unpersuasively that the Supreme Court had in turn reversed the court of appeals. In one of the strongest anti-Carswell statements, former Yale Law School Dean Louis Pollak said that Carswell "presents more slender credentials than any nominee for the Supreme Court put forth in the century,"[25] the statement that provoked Senator Hruska (R-Neb.) to proclaim the need to have mediocrity represented on the Supreme Court.

Prior judicial service may also reveal a judge's ethical sense, a matter which has received much attention in recent years. An indication of the closer attention now paid to matters of ethics was the close questioning of John Paul Stevens about the corporations in which he had held stock or been an officer. (He had resigned as officer when he joined the Seventh Circuit Court of Appeals and later sold most of the securities.) Attention came to be paid to the question of ethics after Justice Fortas resigned upon the disclosure that he had temporarily retained money from Louis Wolfson, a financier in trouble with the government, as part of a contract to be a foundation consultant to Wolfson. This action created the appearance that Fortas was practicing law for Wolfson while on the Court or perhaps—and worse—trying to influence the government's prosecution of Wolfson. The ABA's ethics committee found that Fortas had violated provisions that a judge should avoid "impropriety and the appearance of impropriety" and that a judge should not undertake work inconsistent with his judicial duties.[26]

Ethics was at the heart of the Senate's rejection of Judge Haynsworth's nomination. His not having recused (withdrawn) from a case in which he appeared to have a financial interest was said to be improper, as was his purchase of stock in a company which had been a litigant in a case before him; the purchase was made after the case was decided but before the decision was announced, something Haynsworth attributed to inadvertence. Haynsworth had been cleared of charges of unethical conduct relating to the first matter,[27] but the second matter hurt his cause, particularly in the aftermath of Fortas's conduct.

Doctrinal qualifications may be revealed not only by prior judicial service but also by previous political involvement, which is said to show "judicial temperament" or lack of it; the term is usually left undefined. Quite likely because he had advocated minimum wage and maximum hours legislation, Louis Brandeis was said to lack judicial temperament. Had Hugo Black's membership in the Ku Klux Klan been known before rather than after the confirmation of his nomination, it would have been used against him. However, he probably would have been confirmed anyhow because of his proven support of blacks, his overall liberalism, and the norm of Senate approval of judicial nominations of its own members. In the 1950s John Marshall Harlan was questioned about his membership in the Atlantic Union, which supposedly indicated a diminished commitment to the sovereignty of the United States, and about his attachment to the liberal (Dewey-Brownell) "wing" of the Republican party. Carswell's alleged racism probably produced the defeat of his nomination. John Parker's campaign statement that Negroes shouldn't vote had apparently been an isolated incident and had occurred in the political climate of 1920, but Carswell's racist campaign statement came in the post-World War II period (1948) and was followed by his charter membership in a private white men's club and by his helping to change the Tallahassee, Florida, municipal golf course into a private club to avoid desegregation, something he did while U.S. Attorney. (Carswell's campaign for U.S. senator after he resigned from the Court of Appeals for the Fifth Circuit seemed to confirm the charges of racism.[28])

Prior political activity also produced opposition to the Powell and Rehnquist nominations. Powell's membership on the Richmond, Virginia, School Board and the Virginia State Board of Education led Congress's Black Caucus to oppose him. Complaints about Rehnquist, whose nomination was opposed by the Leadership Conference on Civil Rights and the American Civil Liberties Union, centered on his 1964 opposition to a public accommodations ordinance in Phoenix, Arizona, his Justice Department work on policy concerning electronic surveillance and mass demonstrations, and his defense of Carswell's nomination as the administration's spokesman. The argument that he had only served as the government's lawyer was not accepted by many.

As these examples suggest, a nominee's past political participation is likely to produce opposition by a variety of interest groups—particularly labor and civil rights groups. Those groups have probably been more interested in protecting interests they thought the nominees would injure or in demonstrating their ability to block a nomination than in the particular individual involved, although opposition to Carswell as a person was quite severe.

The American Bar Association's particular stated emphasis has been attention to professional qualifications, as we saw in connection with lower court judges. The organization's relationship to the executive concerning Supreme Court nominations has not been consistent over the last decade. This has been largely a result of the executive's, not the group's, actions. The ABA itself, however, has taken actions which have led to problems. In 1962 it had decided to use only the ratings Qualified and Not Qualified for Supreme Court nominees, who, unlike nominees to lower court positions, were not cleared with the ABA before being announced. Prior to the Haynsworth nomination, the ratings Highly Acceptable, Acceptable, or Not Acceptable were adopted, but after Haynsworth's defeat and before Carswell's nomination the organization reverted to the Qualified/Not Qualified standards. Because people did not know of the change, Carswell's nomination was damaged when he was rated Qualified. (Had people known it was by a majority rather than a unanimous vote, it would have hurt more.) After Carswell's defeat, the ABA again changed it ratings. The new ones were that the nominee met "high standards of integrity, judicial temperament and professional competence," was "not opposed," or was "not qualified."

Probably because of embarrassing information about Haynsworth and Carswell not uncovered by its own investigations, the administration agreed to allow the ABA committee to evaluate not only the person already selected but all those being considered for a Supreme Court position. This arrangement lasted only a short time, however, because the committee rated California Judge Mildred Lillie Not Qualified (by an 11-to-1 vote) and on Little Rock Attorney Herschel Friday six voted Qualified and six Not Opposed; with eight votes needed for a Qualified rating, this division left a Not Opposed rating. Lillie and Friday were dropped from consideration by the president but, in addition to accusing the committee (unfairly) of having "leaked" the nominees' names to the press, he dropped the ABA's prior screening of nominees. The ABA thus knew no more than anyone else (nothing) about the Powell and Rehnquist nominations when they were announced. The vacancy created by Justice Douglas's resignation provided the ABA its desired prior involvement in screening for the first time. Attorney General Edward Levi sent a list of potential nominees to the ABA, which could add to the list. Fifteen names were screened over a two-week period and after interviewing lawyers and judges and reviewing over 200 of his opinions, the ABA gave John Paul Stevens, the ultimate choice, its highest rating.

Supreme Court justices themselves at times get involved in selection politics. Chief Justice William Howard Taft frequently exerted pressure on behalf of people he wanted nominated. Most recently, Chief Justice Warren attempted to exert some pressure on behalf of Arthur Goldberg and Byron White for the Frankfurter and Whittaker vacancies and also on behalf of

Fortas to succeed him as Chief Justice, but he apparently also paid a visit to his old political enemy President Nixon to comment against Justice Stewart's effort to succeed him. The most frequent method used in these efforts by the justices has been a letter of recommendation (34 instances), followed in frequency by personal visits to either the president or the attorney general (20). Next most frequent is a request to the justices from the president—through which they are able to supply information and recommendations, even if the request was not made with that in mind. Intense lobbying is used only infrequently. Efforts in favor of a candidate appear to be less successful than those against one; the success rate in the former category is only 60 percent (46 of 76) while it is 83 percent (15 of 18) in the latter, although other factors might have been determinative even without the judges' intervention.[29]

President and Senate

The president's position in relation to the Senate is stronger for Supreme Court nominations than for those to the lower courts. That position is, however, not absolute, shown by the Senate's successive rejections of two of President Nixon's nominees—the first formal rejection of a nominee since 1930 and the first "double rejection" since 1894. Despite these rejections and the Senate's failure to confirm the nomination of Justice Fortas to be Chief Justice and the nomination of Judge Homer Thornberry which accompanied it (both withdrawn), presidents have done far better in the twentieth century than earlier. Including those of 1894, seven nominations were rejected before 1900 and nineteen others have not been confirmed—for example, withdrawn after the Senate's extended delay in acting on them.[30]

Most nominations sail smoothly through the confirmation process. Usually less than two months is necessary from submission of the nomination to the Senate to final Senate vote, and the time was less than three weeks each for Harry Blackmun and John Paul Stevens. Most of the time is consumed in the period between committee hearings and release of the committee's report; once it receives the nomination, the full Senate votes promptly. Judiciary Committee hearings are perfunctory in most instances, with one-day hearings the pattern in the past, although those for Brandeis occupied *nineteen* days. A majority (9 of 17) of post-1950 hearings have occupied at least two days, but in only three instances were there four or more days of hearings: Thurgood Marshall (4), Haynsworth (8), and Abe Fortas to be Chief Justice (11). Negative votes have occurred in committee on only one-fourth of the nominations and there have been close votes (margins of four or fewer votes) in only three of fifty-two cases in this century: Parker (6–10), Brandeis (10–8), and Haynsworth (10–7).

The nominee's actions during the confirmation process, particularly at these hearings, may have some effect on the outcome. When Harlan Fiske

Stone was under fire from Senator Walsh of Montana, Stone offered to appear before the Judiciary Committee—not the custom at the time. Stone's offer was accepted and he handled himself quite successfully, indicating the importance of the nominee's candor with the committee. Although William Rehnquist provided affidavits to the committee chairman dealing with questions which arose after his hearing, his failure to reappear angered some senators. More damaging was Judge Haynsworth's failure to deal with ethical conduct charges, and Carswell's chances were hurt when he failed to indicate that former Fifth Circuit Chief Judge Tuttle, after saying he had confidence in Carswell and would testify in his behalf, had told Carswell he would no longer testify. Carswell's problem was compounded when the Justice Department failed to arrange a meeting between the nominee and three dubious senators.

The Senate takes no more than two days of debate on all except the most controversial nominations; it used five days each on Fortas (to be Chief Justice) and Rehnquist and seven days on Haynsworth. Until recently the Senate seldom even bothered with roll calls on the nominations, and Senate votes are often unanimous (Blackmun, Stevens) or nearly so (one negative vote on Powell, three on Burger). The rejections, however, have been relatively close: Parker, 41–39, Haynsworth, 55–45, and Carswell, 51–45. In only three twentieth-century confirmed nominations did the negative votes approach one-half the positive vote, and all occurred early in the century: Mahlon Pitney (50–26), Brandeis (47–22), and Charles Evans Hughes to be Chief Justice (52–26). Yet although few recent votes in favor of confirmation have been close (68–26 on Rehnquist was the closest), opposition has increased recently. (Since 1949, seven nominees have received ten or more "no" votes.) Opposition has even occurred when senators were nominated; Hugo Black and Sherman Minton (a Truman appointee) each received sixteen negative votes.

Other factors that may well affect the outcome include the Senate's partisan composition and the timing of the nomination during a president's term. Presidents' nominees have been confirmed 87 percent of the time during the president's first three years in office, but only 65 percent of later nominations have been successful. However, when there is a partisan difference between the Senate and the president, Nixon's situation, in the first three years of his term the president has had only a 64 percent success rate; this falls to a devastatingly low *27 percent* (4 of 15) in the last year. Partisanship's independent effects are notable: the president's success rate is 91 percent when the Senate majority is of his party, only 42 percent when his party is in the minority.[31] The liberalism of senators, rather than their party, has also been said to have a strong effect on their votes in recent nominations —particularly those of Fortas to be Chief Justice, Haynsworth, Carswell, and Rehnquist.[32]

The president cannot control when a vacancy occurs, and even when a nomination is confirmed, opposition may have a carry-over effect. For example, Hoover's nomination of John Parker, which occurred late in Hoover's term, was affected by the earlier controversy over Hughes's nomination to be Chief Justice. Liberal opposition to Warren Burger's appointment as Chief Justice was still simmering when Haynsworth was nominated. That displeasure served to reinforce the deeper liberal and Democratic unhappiness both over the Republicans' having deprived them of the Chief Justiceship when they had blocked the Fortas nomination and over the Nixon administration's not-well-concealed role in helping drive Fortas from the Court. Similarly, when President Ford had to nominate a replacement for Justice Douglas, Ford's earlier unsuccessful attempt as House minority leader to impeach Douglas was in people's minds and Ford might have had considerable difficulty had he named a conservative rather than the moderate Stevens.

Despite this, the president may in the short run be able to affect the timing of the nomination. For example, immediately after the grueling fight over Haynsworth, the Senate, as suggested by one senator, was prepared to confirm anyone who had not raped a small child, in public, recently, so Nixon's delay in sending the Carswell nomination to the Senate misfired. The liberals regained their strength and discovered much negative evidence about Carswell. Rehnquist's nomination, on the other hand, was helped by timing, coming as it did at the end of a session when the necessity of completing normal end-of-session business meant less time for consideration of the nomination, when the Senate was tired from the fight over Earl Butz's nomination to be Secretary of Agriculture, and after substantial effort had been devoted to attacking the anticipated Lillie and Friday nominations.

If the president uses recess appointments to the Supreme Court, as Eisenhower did in three instances (Warren, Brennan, and Potter Stewart), he is likely to produce negative Senate reaction, such as delay in confirming the nomination, in which the Senate engaged with the Stewart nomination to show its displeasure. Yet there are pressures on the president to fill a vacancy on the Court which arises when the Senate is not in session. To wait to make the appointment until the Senate returns might mean injustice to those with pending cases, which might have to be delayed or reargued. Although the next senior justice could "run the show," a vacancy in the Chief Justiceship is particularly serious. Coupled with the fact that *Brown v. Board of Education* had been set for reargument when Chief Justice Vinson died, this probably explains the recess appointment of Earl Warren.

Despite the occasional need to make them, recess appointments cause problems for the president, the Senate, and the Court itself. If the Senate were to convene in special session to consider only the recess nomination, more attention would be focused on a single nomination than it deserved and the Senate's power would increase in relation to the president's. Yet the Senate finds it cannot get necessary information from an already sitting nominee because ethically he cannot respond to questions which touch on cases under consideration by the Court. There is also the possibility that if the appointee is concerned about Senate reaction to his votes, he might "pull his punches" or cases in which he is to write for the Court might be "held" until after his confirmation. (Two important opinions by Justice Brennan were not announced until nine weeks after his confirmation.)

Although often the president does no more—and need do no more—than announce the nomination, not only his timing but also his stance toward the nomination may affect the outcome, particularly if opposition arises. Both Woodrow Wilson's support for Brandeis and Eisenhower's support for Earl Warren (a public statement and a letter to the Judiciary Committee) had positive effects. President Hoover's calling in several Republican senators for discussions to gain support for the Parker nomination was, however, unsuccessful, and Lyndon Johnson's support of Abe Fortas's nomination to be Chief Justice backfired because Fortas was already under attack as a "crony" of Johnson. President Nixon failed to help the Haynsworth nomination with his initial heavy pressure, a special news conference repudiating anti-Haynsworth charges in detail, and a statement that senators should not take a nominee's philsophy into account; Republican senators did not like Nixon's "arm twisting." His attempt at a "low profile" for the Carswell nomination was followed by overreaction when negative information was revealed. In a letter to Senator (later Attorney General) William Saxbe (R-Ohio), Nixon even claimed his right to appoint whom he wanted:

> What is centrally at issue in this nomination is the constitutional responsibility of the President to appoint members of the Court—and whether this responsibility can be frustrated by those who wish to substitute their own philosophy or their own subjective judgment for that of the one person entrusted by the Constitution with the power of appointment. The question arises whether I, as President of the United States, shall be accorded the same right of choice in naming Supreme Court Justices which has been freely accorded to my predecessors of both parties . . .[33]

This statement, challenging the Senate as an institution, was unwise politically and probably lost the nominee some votes.

Off-the-Court Activity

After a nominee to the Supreme Court has been confirmed, his contacts with the president, if he has had contacts earlier, do not cease. If the nominee has been a presidential adviser, that role may continue but is likely to produce criticism, as it did for Brandeis and Wilson, Frankfurter (earlier a member of the Brain Trust) and Franklin Roosevelt, and Abe Fortas and Lyndon Johnson. Even when prior contact has been limited, for example, Richard Nixon's acquaintance with Warren Burger when Burger served in the Justice Department, it may lead to expanded contact between president and judge, particularly for the Chief Justice. Thus Burger apparently did meet on occasion with the president or Attorney General Mitchell, although not, we are told, about cases.

The involvement of justices in noncourt functions has been a source of continuous controversy because through their absence it can produce an evenly divided court, or it can involve a justice in a matter which would later come before the Court.[34] When the Court's workload was not heavy, outside activity did not interfere with the Court's work. Both Chief Justice John Jay and Justice Oliver Ellsworth were involved in treaty negotiations, and Jay, who was also a candidate for governor of New York, served as secretary of state. In the late nineteenth century, Melville Fuller and David Brewer arbitrated a boundary dispute between Venezuela and British Guiana, five justices served on the commission to help resolve the disputed 1876 Hayes-Tilden presidential election, and Stephen Field served on a commission to revise state laws in his native California. Earlier in this century, Charles Evans Hughes served on a commission to determine second-class postal rates. His nomination to be president, which came while he still sat on the Court, was thought by many to injure the Court's nonpartisan image, as was discussion in the 1940s of Justice Douglas's interest in being the Democratic vice-presidential nominee.

More recent "calls" from the president to take on extracourt jobs have encountered more objections because such jobs interfere with disposition of the Court's increased workload and because it has been feared that the Court's independence would be injured if it appeared that justices would readily serve at the president's bidding. During Franklin Roosevelt's presidency, Justice Reed chaired a committee to improve the civil service, Owen Roberts conducted an investigation of Pearl Harbor, and Robert Jackson was chief U.S. prosecutor at the War Crimes Trials at Nuremburg. Jackson's extended absence brought conflict within the Court as well as criticism for thrusting a judge into a simultaneous prosecutorial position and led Chief Justice Stone to turn down Roosevelt's requests to assume other tasks, such as investigating the nation's rubber supply. (During Taft's Chief Justiceship, Stone, a former attorney general, had been willing to serve on the Wickersham Commission investigating crime, but Taft had resisted.) Most recently, Chief Justice Warren, who generally stayed away from most extracourt activity as a matter of propriety, headed the commission to investigate the assassination of President Kennedy, but he did so only after resisting President Johnson's request. Subsequent questions about the Commission's report, including Senate Intelligence Committee revelations that the FBI and CIA withheld information, reinforce the case against this type of extrajudicial activity.

Justices' off-the-court activity has included teaching in law school (Justice Story taught at Harvard and both the first Justice Harlan and Justice Brewer taught at the predecessor to George Washington Law School) and at seminars for judges. The "flap" over Justice Fortas's receiving $15,000, raised by friends, to teach a law school seminar, plus judges' increased public speaking, led the Judicial Conference in 1969 to adopt rules prohibiting the acceptance of fees unless the circuit judicial council approved after determining that the services to be rendered would be in the public interest, were justified by exceptional circumstances, and would not interfere with judicial duties. Such Conference rules, however, do not bind Supreme Court members, although several justices said they would follow them voluntarily.

Drafting legislation affecting the courts and commenting on matters directly affecting the Supreme Court are activities more directly related to the justices' regular tasks. Justice Willis Van Devanter drafted the Judiciary Act of 1915 modifying the Court's appellate jurisdiction and Justice McReynolds, helped by Van Devanter and Justice Day, drafted the 1916 Judiciary Act. Chief Justices William Howard Taft and Charles Evans Hughes appeared at bar association meetings and at the American Law Institute (ALI) to comment on the "state of the judiciary." Such presentations were revived by Chief Justice Warren, who preferred the ALI as his forum because of ABA criticism of the Court's decisions. Chief Justice Burger, in addition to making numerous appearances to stress the importance of court administration,[35] has delivered an annual State of the Judiciary address at ABA meetings. In those speeches he has called for restructuring of the circuits; for elimination of three-judge district courts and of federal diversity of citizenship jurisdiction; and for congressional "court impact statements" indicating how many more judges and supporting personnel would be needed to handle litigation under new statutes.[36] He was joined by most of the justices in responding to the Hruska Commission concerning workload and the preliminary proposal for a National Court of Appeals.

Some justices, going beyond making statements concerning the judiciary, have not been able to resist the temptation to speak out on matters of general public policy. Before the Court's traditions were well established, in the first few years justices were active participants in partisan activity and spoke openly on partisan subjects. From 1810 to the Civil War, judges' extracourt speeches and public letters were less partisan. They became even more quiet after the Civil War, perhaps to help repair the damage from the *Dred Scott* case. At the end of the nineteenth century, when the Court's prestige was again improved, there was an "explosion" of speeches and writing. After the

Depression, the level of oratory again fell and has generally remained low, except perhaps for Justice Douglas's extensive book writing and his statements on conservation and the environment and other justices' bar association activities. Despite the end-of-nineteenth-century outpouring, there seems to have been no relation between the level of the justices' off-court commentary and the Court's prestige, the policy directions of its decisions, or the judges' liberal/conservative tendencies.[37]

This discussion of the off-the-court activity of Supreme Court justices concludes our discussion of the process by which federal judges—in the lower courts as well as the Supreme Court—are chosen and confirmed and of their backgrounds and characteristics. As the conclusion of this part of the book, it also brings to a close our examination of the structure and personnel of the federal judicial system. We can now turn, in the next part, to an analysis of the central part of our subject—the treatment of cases within the federal judicial system, particularly by the Supreme Court itself.

NOTES

1. See Goldberg's statement: "I shall not, Mr. President, conceal the pain with which I leave the Court after three years of service. It has been the richest and most satisfying period of my career." Quoted in Robert Shogan, *A Question of Judgment: The Fortas Case and the Struggle for the Supreme Court* (Indianapolis: Bobbs-Merrill, 1972), p. 108.
2. *American Bar Association Journal* 61 (June 1975): 777.
3. See Henry Abraham, *The Judicial Process*, 2nd ed. (New York: Oxford University Press, 1968), pp. 43–46, for more details on impeachments.
4. Harold W. Chase, *Federal Judges: The Appointing Process* (Minneapolis: University of Minnesota Press, 1972), pp. 36–37. The following section draws heavily on Chase's material.
5. Ibid., p. 23.
6. The best study of the ABA's involvement is Joel Grossman, *Lawyers and Judges: The ABA and the Politics of Judicial Selection* (New York: Wiley, 1965). For a recent thorough discussion of the work of Committee on Federal Judiciary, particularly the procedures it uses in evaluating judicial nominees, see "The Committee on Federal Judiciary: What It Is and How It Works," *American Bar Association Journal* 63 (June 1977):803–7.
7. Chase, *Federal Judges*, pp. 130–31.
8. Ibid., p. 135.
9. See Victor Navasky, *Kennedy Justice* (New York: Atheneum, 1971), pp. 244–76.
10. Sheldon Goldman, "Voting Behavior on the United States Courts of Appeals, 1961–1964," *American Political Science Review* 60 (June 1966): 374–83; Goldman, "Conflict and Consensus in the United States Courts of Appeals," *Wisconsin Law*

Review (1968), pp. 461–82; Goldman, "Conflict in the U.S. Courts of Appeals, 1965–1971: A Quantitative Analysis," *University of Cincinnati Law Review* 42 (1973): 635–58.

11. Stuart Nagel, "Political Party Affiliation and Judges' Decisions," *American Political Science Review* 55 (1961): 843–90; reanalysis of the data is in "Multiple Correlations of Judicial Backgrounds and Decisions," *Florida State University Law Review* 2 (Spring 1974): 258–80.

12. The data which follow are drawn from Sheldon Goldman, "Characteristics of Eisenhower and Kennedy Appointees to the Lower Federal Courts," *Western Political Quarterly* 18 (1965): 755–62, and Goldman, "Judicial Backgrounds, Recruitment and the Party Variable: The Case of the Johnson and Nixon Appointees to the United States District and Appeals Courts," *Arizona State Law Journal* (1974): 211–22.

13. Robert Scigliano, *The Supreme Court and the Presidency* (New York: Free Press, 1971), p. 105.

14. John R. Schmidhauser, *The Supreme Court: Its Politics, Personalities and Procedures* (New York: Holt, Rinehart, and Winston, 1961), pp. 31–32.

15. John R. Schmidhauser, "Judicial Behavior and the Sectional Crisis of 1837–1860," *Journal of Politics* 4 (November 1961): 615–40.

16. Stuart Nagel, "Comparing Elected and Appointed Judicial Systems," Sage Professional Papers #04–001 (Beverly Hills, Calif.: Sage Publications, 1973), p. 25.

17. John R. Schmidhauser, "*Stare Decisis*, Dissent and the Background of the Justices," *University of Toronto Law Review* 14 (May 1962): 194–212.

18. David W. Rohde and Harold J. Spaeth, *Supreme Court Decision-Making* (San Francisco: W. H. Freeman, 1975), pp. 107–108.

19. Scigliano, *The Supreme Court and the Presidency*, p. 146. His full treatment of the subject is on pp. 125–28.

20. Charles M. Lamb, "The Making of a Chief Justice: Warren Burger on Criminal Procedure, 1956–1969," *Cornell Law Review* 60 (June 1975): 786.

21. Ibid., p. 756.

22. Charles M. Lamb, "Exploring the Conservatism of Federal Appeals Court Judges," *Indiana Law Journal* 51 (Winter 1976): 257–79.

23. *Vlandis v. Kline*, 414 U.S. 441 at 463 (1973).

24. This examination draws on Joel B. Grossman and Stephen L. Wasby, "Haynsworth and Parker: History Does Live Again," *South Carolina Law Review* 23 (1971): 345–59, and Grossman and Wasby, "The Senate and Supreme Court Nominations: Some Reflections," *Duke Law Journal* (August 1972):557–91.

25. *Congressional Record*, 116 (1970), p. 2860.

26. See Shogan, *A Question of Judgment*.

27. For a correction and addition to Grossman and Wasby based on correspondence with Judge Haynsworth, see Wasby, *Continuity and Change: From the Warren Court to the Burger Court* (Pacific Palisades, Calif.: Goodyear Publishing, 1976), pp. 27–28, note.

28. His arrest in 1976 for homosexual activities, had it come earlier, would have "finished him off" completely. (He was convicted of battery.) Fortas has been criticized for having helped "hush up" a story involving such an arrest of one of President Johnson's assistants.

29. Henry J. Abraham and Bruce Allen Murphy, "The Influence of Sitting and Retired Justices on Presidential Supreme Court Nominations," *Hastings Constitutional Law Quarterly* 3 (Winter 1976): 37–63.

30. For this earlier period, see Abraham, *The Judicial Process*, pp. 82–85. This section is based on Grossman and Wasby, "The Senate and Supreme Court Nominations."

31. See Scigliano, *The Supreme Court and the Presidency*, pp. 146–47.

32. Rohde and Spaeth, *Supreme Court Decision-Making*, pp. 105–106.

33. Richard M. Nixon to William Saxbe, March 31, 1970, *Congressional Record* 116 (1970): p. 10158.

34. Russell Wheeler, "Extra-Judicial Activities of the Early Supreme Court," *Supreme Court Review* 1973, edited by Philip Kurland (Chicago: University of Chicago Press, 1973), pp. 122–58.

35. See Arthur R. Landever, "Chief Justice Burger and Extra-Case Activism," *Journal of Public Law* 20 (1971): 523–41. See also Burger's magazine article, "The Chief Justice Talks About the Court," *Reader's Digest* 102 (February 1973): 99ff.

36. See "The State of the Federal Judiciary—1972," *American Bar Association Journal* 58 (October 1972): 1049–53; "Report on the Federal Judicial Branch—1973," *American Bar Association Journal* 59 (October 1973): 1125–30; "Report on the Federal Judicial Branch—1974," *American Bar Association Journal* 60 (October 1974): 1193–98; "The State of the Judiciary—1975," *American Bar Association Journal* 61 (April 1975): 439–43; "Annual Report on the State of the Judiciary," *American Bar Association Journal* 62 (April 1976): 443–46.

37. Alan F. Westin, "Out-of-Court Commentary by United States Supreme Court Justices, 1789–1962: Of Free Speech and Judicial Lockjaw," *An Autobiography of the Supreme Court: Off-the-Bench Commentary by the Justices* (New York: Macmillan, 1963), pp. 28–29.

PART THREE | THE TREATMENT OF CASES

In this part of the book, we turn to the way in which cases are treated in the federal judiciary and particularly by the Supreme Court. Very few cases filed in the federal or state trial courts ever get to the United States Supreme Court, even in the form of a request for review. Cases once filed are often not pursued, many civil cases are settled out of court, and most criminal cases—as much as 85–90 percent in some jurisdictions—are resolved with a plea of guilty. Of those which reach the trial stage, many never proceed further. The relatively few taken to the first appellate stage usually are not pursued beyond that point. State cases which do move further are first appealed through the state court system. The great majority of state cases are within the Supreme Court's certiorari jurisdiction, allowing the Court to reject the great majority of them; the Court disposes of most cases from the federal appeals courts similarly.

The formal paths along which cases move are clearly delineated. However, before people may begin to move cases along those paths, they must satisfy the Supreme Court's requirements as to who can have access to the courts and which cases can be brought to court. The Supreme Court sets such rules primarily for the federal courts. While the justices cannot tell state courts which cases to hear, they are not bound to accept state cases for review if those access requirements are not met. Other Supreme Court access rules are intended to keep cases in the state court systems or at least to ensure that an attempt is made at complete treatment there before the cases are brought to the federal courts. Within the constraints imposed by all these rules, lawyers—and the groups behind some of the lawyers—make their decisions to take cases on appeal and try to shape issues for appellate presentation. Lawyers' actions limit the justices' freedom of movement but do so only in part. The justices themselves, even if they must wait for cases to come to them, can have a considerable effect on the issues they decide. Their policy decisions affect cases which are subsequently initiated in the lower courts, adding to the effect of their procedural doctrine.

The Supreme Court uses certain regular procedures to dispose of its business, particularly at the stage when the Court is asked to review cases, with the Court denying most petitions for review. Thus while judicial requirements have eliminated many potential cases at earlier stages, the Court itself removes more. This "skimming off" of most cases and the justices' disposition of a number of other cases through brief orders (summary dispositions) allows the Court to concentrate on those it considers most important. In such cases, the Court hears oral argument and the justices write opinions. The alignments or blocs within the Court, through which the justices' attitudes can be seen, are most visible there. Those cases are the vehicles through which the Court makes its most important contributions to policy and the ones to which its attentive publics pay most heed.

In Chapter 5, the rules the Supreme Court has developed to control entry of cases to the courts and its doctrine concerning the relations between federal and state courts are discussed. This is followed by a preliminary exploration of groups' involvement in cases and lawyers' decisions to appeal. Chapter 6 begins with a look at the size and content of the Supreme Court's docket and continues with discussion of the Court's disposition of petitions for review and its summary disposition of cases. Chapter 7, which is devoted to the Court's "full-dress" treatment of major cases, includes a description of the justices' basic procedures for reaching decisions in those cases, with particular attention given to oral argument, the Chief Justice's opinion-assignment practices, and the release of opinions. Agreement and disagreement among the justices and their alignments in deciding cases are also analyzed in that chapter.

5 | UP TO THE SUPREME COURT

ACCESS TO THE COURTS

Many of the Court-developed rules concerning the entry of cases into the courts, part of the crucial process by which cases are "weeded out" of the judicial system, revolve around the constitutional requirement—in Article III—that courts may deal only with "cases or controversies." This is a corollary of the idea that the courts are legal institutions acting within a specific procedural framework. The rules on access include requirements that the parties to a case be adverse to each other and that the case not have been "manufactured" to get it into court, that someone bringing a case have standing to sue, that the case be both "ripe" and not moot, and that it not entail a "political question." Questions about such rules are most likely to arise when someone wishes to challenge a law and to obtain a declaration of his or her rights prior to enforcement of the law against that person; it is easier to challenge government restrictions already imposed and it is clear that one may question the validity of a law as a defense to a criminal charge.

The Court's doctrines on access to the courts have changed over time. The Warren Court generally made access to the courts easier, while the Burger Court has been considerably more restrictive. The Burger Court similarly developed barriers for people trying to get their cases from the state courts to the federal courts. In making it more difficult for people to deal with certain types of social problems through the legal system, the Burger Court has forced the judicial process back toward a narrower adversary model which it has been the purpose of the access rules to provide.

Advisory Opinions and Feigned Controversies

The Supreme Court's most basic rule, settled early in the Court's history, is that it will decide questions of law only when they are presented in the fact context of a particular lawsuit. In other words, the Court will not issue advisory opinions—opinions on abstract legal questions. Facts, say the judges, give the law its meaning. For example, a law might appear to be constitutional "on its face" but be applied in an improper manner, or a law the validity of which is at first dubious might have been applied—and interpreted by the lower courts—in a way to ensure its validity. The rule against advisory opinions was established when President Washington asked the Court for advice on some treaty questions; the justices responded through Chief Justice Jay that they would not do so. Early in the present century, the Court amplified the rule. Congress had passed a law allowing the Cherokee Indians to file suit to test a statute, but in *Muskrat v. United States* the Court said that Congress was seeking a ruling on the law's constitutionality without the presence of a live controversy, therefore in effect asking for an advisory opinion, and the justices would not allow the case to be heard.

Some informal exceptions to the ban on advisory opinions have, however, occurred. In 1822 President Monroe asked several members of the Court about the legality of federal "internal improvement." Justice William Johnson made a general if oblique reply for himself and several other justices, indicating their position that governmental construction of military roads and postroads was constitutional. Certainly Chief Justice Hughes's comment on behalf of the Court as to Roosevelt's Court-packing proposal that the "one Supreme Court" called for by the Constitution could not sit in divisions could be called an advisory opinion. At other times, individual justices have provided the president with legal advice. For example, Abe Fortas helped Lyndon Johnson by drawing up legislation. (Chief Justice Taft apparently advised Calvin Coolidge, when the latter became president, that the best thing he could do in being president was nothing.)

In the course of deciding cases, justices also comment on matters on which the Court is not ruling, making their statements somewhat like advisory opinions. Justices writing concurring or dissenting opinions are more likely to make statements of this sort. When the author of the Court's majority opinion does this, the statements are called *obiter dictum*, that is, language beyond the "holding" or basic rule of the case. (Sometimes it is difficult to tell what is *obiter* and what is not.) The Court as a whole has also been charged with issuing advisory opinions when it goes beyond the specific facts of a case to anticipate general questions of policy not yet directly presented to it.[1] Those making such a charge fail to recognize that because the Supreme Court is the nation's highest court, its decisions are intended to have a broad reach, to decide more than the particulars of the dispute between the parties to the individual case. The Court accepts for review cases intended to be representative of larger issues, something it indicates when it decides several cases on the same topic at one time.

In order to avoid handing down advisory opinions, the Court insists that litigation not be "made up" simply to obtain a ruling on some matter which the parties would like to see settled, particularly to the detriment of a third party not involved in a case. Thus the parties in a case must be adverse to each other. This rule against feigned controversies is aimed at collusive litigation. It does not mean that the litigants have to be "at each other's throats" but only that their interests be opposed. One of the Court's earliest statements of the rule was in a case, *Lord v. Veazie,* in which a stockholder, claiming ownership of river navigational rights also claimed by someone else, sold some stock based on his own presumed control of the rights. By prearrangement, the stock purchaser, a relative of the seller, sued the seller to determine ownership of the river rights. Both parties obviously wanted the same result, which they were trying to ensure without the third party's involvement. The Supreme Court ruled that no court should hear such a lawsuit.[2]

Probably the most obvious instance in which the Court ignored its own rule on feigned controversies came during the New Deal in *Carter v. Carter Coal Company* when the president of a coal company sued his own company to prevent it from complying with the Bituminous Coal Act. The government's opportunity to defend its own statute was severely limited in such a case because it was not a party. The fact that the president of the company and the company itself did not have interests adverse to each other was ignored by the Court. To avoid repetition of such a situation impinging on federal law, Congress soon passed a law which automatically made the United States a party to any case in which the validity of a federal statute was challenged.

Standing

Closely related to the need for adverse parties is the requirement that the person bringing a case have standing, that is, be the appropriate one to do so. If no one has proper standing, the court lacks jurisdiction—that is, legal authority—to hear the case. As it may not be possible for anyone to meet standing requirements to challenge a law, regulation, or practice, the possibility exists that unconstitutional rules will remain on the books because no one can "get at them." The rules on standing are largely judge-made, although Congress has at times indicated who may or may not bring certain types of cases in the federal courts. For example, the Court has said that the law establishing the National Railroad Passenger Corporation (Amtrak) allows only the attorney general to enforce its provisions and thus bars a railroad passenger group from protesting passenger train discontinuances. Similarly, a labor union member may intervene in a Landrum-Griffin Act election case but may do so only to discuss those claims presented in the secretary of labor's complaint. The Court has also ruled, however, that because Congress meant to define standing broadly in the 1968 Open Housing Act, white tenants of an apartment complex had standing to complain about discrimination against blacks.[3]

The basic rule on standing is that the litigant must show injury to himself or herself; thus a doctor could not challenge anticontraceptive statutes solely on the basis that they would injure his patients.[4] At times, however, the rights of others may be asserted to reinforce one's own claim, particularly where there are obstacles to others' asserting their own rights and where the party in court would assert those rights effectively. Thus a white woman, sued for damages for selling her house to blacks in violation of restrictive covenants, was granted standing to assert blacks' rights to purchase housing; a married man charged with violating a law which prohibited supplying contraceptives to single people but which did not punish the recipients was able to assert their rights; and doctors were allowed to assert the rights of Medicaid patients who might seek abortions in the doctors' suit to recover payment.[5] Standing has also been granted to whites to challenge the exclusion of blacks from juries and to men to challenge the exclusion of women on the basis that all have a right to a jury which is a properly drawn cross-section of the community.[6]

During the late 1960s the Court relaxed the rules on standing. The class of people said to be injured was expanded and federal taxpayers were granted standing to sue in some instances. In expanding the scope of those "injured,"[7] the Court also recognized in *Sierra Club v. Morton* that injury to "aesthetic and environmental well-being" was among the interests which could provide a basis for standing. However, the majority made clear that to obtain standing a group would have to do more than state a general interest in the environment; its members would have to be among those injured. Justice Blackmun, who would have allowed standing without the personal injury requirements, asked in dissent, in what was hardly a position of judicial self-restraint: "Must our law be so rigid and our procedural concepts so inflexible that we render ourselves helpless when the existing methods and the traditional concepts do not quite fit and do not prove to be entirely adequate for new issues?" Later, when a personal injury claim was added to claims of injury to the environment, the Court was willing to grant standing.[8]

Prior to 1968, federal taxpayers alleging that money was being used for an unconstitutional project could not use their taxpayer status to establish standing to sue the government.[9] In 1968 the court ruled in *Flast v. Cohen*, a case brought to test the 1965 Elementary and Secondary Education Act (ESEA) provisions for assistance to parochial schools, that a federal taxpayer could have standing to sue the government if the challenged program were alleged to violate a specific constitutional prohibition, such as the First Amendment's "establishment of religion" clause. However, taxpayer attacks based on more general constitutional provisions were still not to be allowed. The Burger Court reinforced this position by saying there was insufficient connection (or "nexus") between being a taxpayer and Congress's failure to require the CIA to produce detailed reports of its expenditures to allow the taxpayer standing, even though the plaintiff asserted agency failure to comply with the Statements and Accounts Clause (Art. I, Sec. 9, cl. 7). Justice Powell emphasized that under our system of government, such claims should be brought to Congress rather than to the courts, and Chief Justice Burger stressed that taxpayers could not use federal courts to assert general citizen or voter grievances. Burger's position was promptly applied to deny standing to armed service reserve officers trying to prevent congressmen from holding reserve commissions as a violation of the Incompatability Clause (no executive official should be a member of Congress).[10]

The Burger Court's recent restrained view of standing has been clearest with respect to civil rights claims. When blacks and whites in Cairo, Illinois, sued to obtain relief from racially discriminatory bonding, sentencing, and jury fee practices, in *O'Shea v. Littleton,* the Court denied standing because the plaintiffs had not shown either that they had already been injured or that they could suffer future continuing injury from the practices. Similarly, in *Rizzo v. Goode,* a suit against Philadelphia's mayor and police commissioner over improper police practices against minorities and mishandling of complaints, the Supreme Court overruled a district judge who had proposed a comprehensive plan for handling civilian complaints. The Supreme Court said the complainants lacked standing to bring about an overhauling of police disciplinary procedures because their claim rested on what a small, unnamed minority of officers might do in the future and on incidents which might happen to others if the police procedures were not changed. The Court said the controversy was too generalized to be decided by judges, an indication that standing rules and other elements of the "case or controversy" question to be discussed below often appear in the same case.

When racial and ethnic minority central-city residents of low and moderate incomes in Rochester, New York, sued to invalidate a suburb's zoning ordinances as excluding the poor, their claims were also turned aside on the basis of standing in *Warth v. Seldin.* A nonprofit corporation seeking to alleviate their housing shortage, central-city taxpayers, and a home builders' association also tried to join the suit, but the Court said neither they nor the principal complainants had a "case or controversy" with the suburb. To sue they would have to indicate specifically how ordinances would harm them, for example, that without the zoning practices they would have been able to purchase or lease homes in the suburb or that the suburb's actions produced an increase in the central city's taxes. To gain standing, they would also have to state how judges could protect their rights. The Court's ruling and Justice Powell's comment that "the economics of the housing market," not the ordinances, prevented plaintiffs' movement to the suburbs caused Justices Douglas and Brennan to say that the Court was making it virtually impossible for anyone to challenge the suburb's actions and was showing "indefensible hostility" to the poor. Douglas, asserting that "standing has become a barrier to access to the federal courts," assumed an "activist" posture in arguing for lowering "the technical barriers" to cases so that courts could deal with the "festering sores" in our society.[11] (The following term, in *Village of Arlington Heights v. Metropolitan Housing Development Corp.*, the Court did grant standing to a developer to test the denial of a requested rezoning of a tract the developer wished to use for low and moderate income housing, and then found no racial discrimination in the denial.)

The Cairo, Philadelphia, and Rochester suits were "class actions" brought by one or more named individuals for themselves and "all others similarly situated." Class actions spread costs for the litigants and avoid repetitious litigation, thus saving time for the courts. The Court has, however, limited their use through statutory interpretation. The first limitation came in *Snyder v. Harris*. A consumer suing a utility for not following its tariffs (rates) had brought a class action on behalf of all those served by the utility in order to have the $10,000 claim necessary to get into federal court in a diversity of citizenship case. (The dollar amount does not apply in civil rights cases.) The Court turned him aside, ruling he could not join his claim with those of others to reach the $10,000 figure. Four years later, in *Zahn v. International Paper Co.*, where lakefront property owners were suing a paper company for industrial pollution of the lake, the Court ruled that the $10,000 amount had to be satisfied by each and every member of the class. If not all had $10,000 claims, those who did could not proceed as a class. (In a related case, the State of Hawaii was denied standing to sue oil companies on behalf of its citizens for antitrust law violations; the state could sue to protect its own direct interests, but consumers would have to bring their own suits.[12])

The Court imposed other barriers to class actions in the complicated *Eisen* litigation. A buyer of small numbers of shares (odd lots) on the New York Stock Exchange filed a class action to challenge antitrust and securities law violations. Members of a class are supposed to be notified so they can withdraw if they wish; in this case, the district judge had allowed the plaintiff to notify only a sample of members of the potential class and, because he thought the plaintiff likely to win the case, ordered the defendant to pay 90 percent of the notification costs—which would have been $200,-000 if all members of the class had to be notified. The Supreme Court, however, ruled that the petitioner(s) must assume all the notice costs and that notice must be sent to all class members identifiable through reasonable effort.

The relationship between the justices' ideology and these procedural rules, perhaps clear from these cases which made it more difficult for complaints to be brought to court, was even more clear when the majority ignored class action problems when it wanted to decide the underlying issue. In *Sosna v. Iowa*, a woman—the only named plaintiff—had brought a class action challenging requirements that one live in the state for a certain time in order to obtain a divorce. By the time the case reached the Supreme Court, the woman had received a divorce elsewhere, eliminating her "case or controversy." Although there were no other named plaintiffs, the Court said that because the district judge had certified it as a class action the case was still alive and then sustained the residence requirement, clearly a conservative result.

Like the rules on standing and class actions, rules on the award of attorneys' fees can have a marked effect on access to the court. Particularly affected are interest groups serving as "private attorneys general" to enforce existing law or bringing broad "public interest" challenges like environmental cases where no single individual or small set of individuals has suffered substantial economic injury of the type which could be compensated through an award of damages. Although in 1968 the Court had ruled that attorneys' fees could be awarded under Title II of the Civil Rights Act of 1964,[13] in 1975 the Court denied them in the *Alyeska Pipeline* environmental litigation because Congress had not specifically provided for them in that type of case. Many "public interest" groups felt that ruling would severely hinder their ability to bring and pursue cases, but Congress's action in passing the Civil Rights Attorneys Fees Act in 1976 alleviated at least some of that concern.

Mootness and Ripeness

The divorce residency case shows another aspect of the "case or controversy" requirements, that a case not be moot, that is, that a dispute not be completed or too late for judges to apply a remedy. In the 1960s the Supreme Court altered its rules on mootness to make it easier for courts to decide issues, particularly where they were quite likely to arise again. Under older definitions of mootness, challenges to election rules, particularly those excluding candidates, were usually moot because an election had normally taken place before a challenge could reach the appellate courts. However, the Court began to rule in some cases where the issue was likely to be repeated in later elections.[14] A prisoner's habeas corpus challenge to his conviction was also ruled not moot, although he had been released from prison because the disabilities attached to the conviction would still remain.[15] In *Roe v. Wade*, the Court's principal abortion case, Justice Blackmun stated that if traditional mootness rules were applied, questions related to pregnancy could never be reviewed on appeal because the pregnancy would be completed. He said, however, that the issues should be decided because pregnancy was likely to recur, both for the woman challenging antiabortion laws and for other women as well, and he ruled that such matters, "capable of repetition yet evading review," should be dealt with.

A case cannot be moot if judges are to rule on it, but it also cannot be "premature" or " unripe"; a person bringing a premature case lacks standing. The concept of ripeness, on which there are few clear rules, is closely related to the idea that courts will not deal with abstract or speculative matters. For example, the Court said in *Poe v. Ullman* that although a prosecutor said he would prosecute those violating a state anticontraceptive law, a challenge to the law was not ripe because there had been no prosecution under the law for twenty years and contraceptives were widely available. Justice Frankfurter said that the Court did not make declarations of people's rights when it was not necessary to do so. (When a doctor opened a birth control clinic and was arrested, the Court did invalidate the law.[16]) Indicating the relationship of doctrines on prematurity and standing, the Court also refused in 1972 to hear a challenge to Army surveillance of civilian political activity, saying in *Laird v. Tatum* that it was unripe because only a subjective "chill" to First Amendment rights was involved, not actual injury.

The doctrine that someone should exhaust available remedies before taking a case further is a part of prematurity. One must make full use of state court remedies before using federal courts to challenge convictions and must avail oneself of appeals within an administrative body before taking matters to court. For example, the Court had ruled that challenges to courts-martial cannot be heard by civilian courts until military courts have completed their review, even where the military's jurisdiction to hold the court-martial is the subject of the challenge.[17] The exhaustion doctrine in areas like welfare may exhaust the individual rather than the remedies and has had a particularly harsh effect on those denied conscientious objector status by their draft boards, where, as a result of statutes sustained by the Court, review is available only through a challenge to a criminal indictment for refusing induction.

"Political Questions"

The most ambiguous facet of the "case or controversy" requirement is the "political question" doctrine that courts will not decide questions which the Constitution indicates the other branches of government should handle or which the judges themselves think are more appropriately handled there. The Court has placed a variety of questions within the scope of this doctrine. These have included whether a state may "de-ratify" a constitutional amendment after initially ratifying it; whether state constitutions may provide for recall of judges; which of two "regimes" in a particular state was the legitimate one; and whether or not a president should enforce certain laws.[18] The Warren Court drastically reduced the content of "political questions." First, although for many years reapportionment had been thought to be a "political question," in 1962 the Court said in *Baker v. Carr* that complaints about malapportioned legislative districts could be heard by the courts. Then, although legislatures had been allowed to judge the qualifications of their own members without judicial interference, in *Bond v. Floyd* the Court overturned the Georgia legislature's refusal to seat Julian Bond because of his antiwar statements and in 1968 ruled that Congress should seat Congressman Adam Clayton Powell (D-N.Y.). The ruling in *Powell v. MacCormack* came close to rejecting the "political question" doctrine entirely.

The doctrine is, however, still applied in some situations. For example, when Kent State University student government officers challenged the Ohio National Guard's riot control rules, the Court said that to provide the requested remedy would involve federal judges in constant supervision of the executive branch and thus judges could not rule on the case.[19] And issues involving our relations with foreign nations, such as the validity of treaties under international law and our recognition of foreign governments, are still considered to be "political questions" so that the courts will not interfere with the executive branch's decisions. The federal courts often used the political question doctrine as the basis for refusing to decide the validity of the Vietnam War under either our own Constitution or international law.[20] Another aspect of the "political question" doctrine in the foreign affairs area had been the "act of state" doctrine, under which our courts will not question other nations' acts done within their own territories. The Court has, however, been badly divided as to how to apply the doctrine, particularly when the State Department has said it need not do so. A number of justices feel that it is inappropriate to accept the executive's word instead of having the judges make their own determination.[21]

RELATIONS BETWEEN FEDERAL AND STATE COURTS

Our federal system, in which both the national government and the states have substantial authority, requires the Supreme Court to define national judicial power so as not to undermine the states' ability to deal with problems. Because jurisdictional lines are not clear, many issues—not only civil rights questions but also more mundane matters such as workmen's injuries —often seem subject to decision by both federal and state courts, and the

Supreme Court must decide which set of courts will rule. (Another issue which has arisen frequently, particularly in recent years, is the question whether state courts, Indian trial courts, or federal courts should have jurisdiction over cases concerning Indians.) More, however, is involved than the competing power of judicial systems. The basic power of state governments as a whole to carry out their work unhindered by federal judges is what is primarily at issue. One relatively new mechanism for reducing friction between federal courts and state courts is the federal-state judicial council. Such councils, now found in most states, are composed of federal and state judges who meet to discuss and try to resolve common judicial problems such as the handling of habeas corpus petitions.

The Supreme Court has also developed an important set of rules covering federal court–state court relations. Adopted to reduce friction between national and state governments, the rules cover use of habeas corpus petitions in federal courts to challenge state convictions, federal court declaratory and injunctive relief against state laws, and removal of cases from state to federal courts. The rules both limit access to the judicial system and make it more difficult to challenge the actions of state legislatures and state administrators. The rules have as their underlying rationale the idea that state courts should decide as much as possible before claimants may bring cases into the federal court system and that, when such claims are brought into that system, review of state court decisions should be limited.

Foremost among the rules to eliminate federal judicial interference with the states are those restricting the filing of federal cases against states and state officials. The Eleventh Amendment, ratified in 1798 to overturn an early Supreme Court ruling, prevents suits against a state by citizens of another state. States are further protected from suit by the common law rule of sovereign immunity, which prevents citizens of a state from suing their own state without its consent. The Supreme Court has recently combined this idea with the Eleventh Amendment by holding that the amendment bars suits by citizens of a state against that state where the payment of funds from the treasury would be necessary to pay for past actions. Thus, when a state had improperly withheld welfare benefits, the Court said in *Edelman v. Jordan* that although injunctions could be sought to prevent future state misbehavior, people could not sue for retroactive payment of the withheld benefits without the state's consent. The Court also said that the state's participation in the congressionally enacted welfare system did not provide the necessary consent.

Although it is clear that state and local units of government cannot be sued for damages,[22] government officials may be sued in some circumstances. Thus if state officials violate the U.S. Constitution, their immunity from suit is removed so that they are personally liable for damages. However, in so ruling in *Scheuer v. Rhodes*, one of the Kent State cases, the Court

> The judicial power of the United States shall not be construed to extend to any suit in law or equity, commenced or prosecuted against one of the United States by citizens of another state, or by citizens or subjects of any foreign state. (Eleventh Amendment)

said that the higher an official, the more discretion he is to be allowed, so that high state officials would have what amounts to executive immunity. The Court reinforced its basic ruling on liability by saying in *Wood. v. Strickland* that school board members, superintendents, and school principals would be personally liable if they violated the rights of students either maliciously or knowingly or in a situation where they should have known they were doing so.

In 1976, however, the Court handed down two rulings which closed off the possibility of damage suits against law enforcement officials. In *Imbler v. Pachtman*, the Court said that a state prosecutor acting within the scope of his duties was absolutely immune from suit under the civil rights statutes, even when a claim that he had knowingly used false evidence and suppressed material evidence at a trial had resulted in reversal of a defendant's conviction. Without absolute immunity, the prosecutor would have to spend much of his time defending against lawsuits instead of carrying out his assigned duties, said Justice Powell. In the other case, *Paul v. Davis*, the majority of the justices said that no federal civil rights claim could be based on defamation of a person's reputation; the ruling came in a case brought by a person listed in a police chief's "flyer" as "active" in shoplifting when his one arrest had resulted in a dismissed case.

Most other rules in this area deal with where a suit should be brought. This includes the states' allowing diversity of citizenship cases to be brought in federal rather than state court when at least $10,000 is involved. The federal courts' application of federal rather than state law in such cases has caused substantial friction within the federal system. In 1842 in *Swift v. Tyson* the Supreme Court ruled that the federal courts could establish their own common law in diversity cases. The Supreme Court did not reverse itself until 1938 in *Erie Railroad Co. v. Tompkins* when it declared that its earlier decision was unconstitutional, that there was no federal common law, and that the federal courts should apply the laws of the states in diversity litigation. Despite *Erie*, there is still the equivalent of a federal common law of procedure because the federal courts have to have a set of rules as to which state's law to apply when different results would occur (the "outcome determinative" test), but state law now has much greater authority in such suits. In large measure to reduce federal court caseload, Chief Justice Burger would eliminate diversity jurisdiction altogether. His argument is that it is simply no longer true that state judges are less fair to an out-of-state litigant than a federal judge would be.

Other rules allow a case started in one system to be shifted ("removed") to the other. During the 1960s, civil rights workers in the South, arguing that they were being arrested and tried in order to harass them for exercising their rights, tried to have their cases transferred to federal courts. However, the Supreme Court, sensitive to the problem of taking cases away from the state courts, took a narrow view of the federal removal statute. Removal could occur, the Supreme Court said, only when there was a statutorily protected federal right involving racial equality, such as the right to service in a place of public accommodation after passage of the 1964 Civil Rights Act, and a state law or rule prevented the protection of rights in a state trial; an allegation of an unfair trial was an insufficient basis for removal, as errors at trial could be corrected on appeal.[23] The Court stood by this limited interpretation when it ruled in 1975 that those charged with conspiracy and boycott, although they claimed protection by the 1968 Civil Rights Act provision forbidding interference with civil rights, could not remove their trials to federal court.[24]

If the removal rulings are based on statutory interpretations, there is also a judge-made doctrine known as abstention for dealing with potential—and real—conflict between federal and state courts. Under this doctrine, federal courts whenever possible will "stay their hand" until state courts have ruled. Where state law is subject to more than one interpretation and has not been previously interpreted by state courts, cases may be returned to state courts for such interpretation, which might "avoid or significantly modify the federal questions" in the case.[25] In connection with this problem, some state courts have provided rules for prompt interpretation of unanswered state law questions when those questions are certified to them by the federal courts. The Court will also return criminal cases to state courts for rulings on unanswered questions so that the adequacy of independent state grounds which might support a conviction may be ascertained.[26] While the Supreme Court does review state court decisions on federal constitutional questions, it seldom interferes with those thought to rest on adequate state law grounds. Thus if state courts carefully base their rulings solely on their own constitutions, they can both avoid review and establish rules different from those the U.S. Supreme Court would require under the national Constitution, for example, higher criminal justice standards for state officials.

The federal courts do not invariably abstain from deciding cases when state courts may also be involved. Recently Justice Brennan said that abstention was "the exception, not the rule" because federal courts could not abdicate their duty to decide cases unless "an important countervailing interest" would be served. Abstention was appropriate, he said, in cases presenting a federal constitutional issue which might be mooted or presented in a different posture by a state court determination of pertinent state law ...

where there have been presented difficult questions of state law bearing on policy problems of substantial public import whose importance transcends the result in the case at bar ...

where, absent bad faith, harassment, or a patently invalid state statute, federal jurisdiction has been invoked for the purpose of restraining state criminal proceedings.

Yet even when abstention was not applicable, "wise judicial administration" —such as avoidance of piecemeal litigation of legal questions—might make it best for a federal court to leave matters to the state courts particularly when the federal court was several hundred miles from the dispute and state court proceedings had been initiated first.[27]

Closely related to abstention is the question of whether state court action is "final" so that appeal to the Supreme Court is proper. The Supreme Court does not wish to accept an appeal from a state court in a case in which only some questions have been answered when other proceedings are still pending, because a Supreme Court ruling at that stage may mean interference in state proceedings. On the other hand, not accepting the appeal may mean needless continuation of the trial and thus injury to the appellants if resolution of an issue would terminate the proceedings. On this question of "finality," the Court takes a pragmatic approach despite complaints that the number of exceptions to the finality rule has become larger.[28]

A preference that all claims in a case be heard by either state or federal courts has been shown by the Supreme Court. At times, such a preference may lead to all claims being decided in federal rather than state courts, the reverse of the result produced by abstention. When state matters are raised in federal court along with an adequate federal claim, they may be heard by the federal judge under the doctrine of "pendent jurisdiction." If the federal court then follows normal rules of self-restraint and avoids federal constitutional questions, it may decide the state claim first, thus ironically not resolving the federal questions.[29] Despite the preference for "all claims in one court," the idea is not followed consistently. Eliminating those without $10,000 claims from diversity of citizenship class actions means that the same legal questions may be argued in federal court, for the others, with potentially conflicting results. Similarly, prisoners' claims of different types but based on the same facts, heard simultaneously in federal and state courts, may lead to inconsistent results and certainly produce duplicative litigation.[30]

The basic rationale for abstention, noninterference in state judicial proceedings, can also be seen in the Court's unwillingness to have the federal judiciary regularly supervise state judges. The Cairo case challenging state judges' allegedly discriminatory practices was rejected in large measure because an injunction, the requested remedy, would have produced "an ongoing federal [court] audit of state criminal proceedings" which would lead to "intrusive and unworkable" intervention. This was to be avoided particularly where other remedies, such as prosecution of the judges for violation of federal criminal rights statutes, might be available.

The federal courts' role in hearing habeus corpus petitions from convicted state prisoners challenging their convictions and requests for injunctions against state prosecutions has been the subject of particular controversy in recent years. A substantial increase in habeas corpus petitions in the federal courts was produced by the Warren Court ruling that, contrary to past law, such petitions would not be barred by existence of an adequate and independent basis in state law for affirming a conviction. Furthermore, as long as a prisoner had not earlier deliberately failed to use remedies provided by the state for challenging a conviction and those remedies were no longer available, he need not have first exhausted them before bringing the federal court habeas corpus action. However, the Court later reinforced the basic "exhaustion" requirment by saying a federal habeas petition could not be filed until a convicted person's specific claim that his rights had been violated, for example, by a defective indictment, had been brought before the state courts.[31] The Burger Court imposed limitations on the use of federal habeas petitions, particularly when it ruled in 1976 in *Stone v. Powell* that search and seizure claims already reviewed by a state court could not be raised again in federal court. Earlier rulings making it extremely difficult for a counseled defendant to challenge matters occurring prior to a guilty plea also were reinforced by a decision that a defendant could not challenge the composition of a grand jury through federal habeas when he had failed to raise the issue before his trial unless he showed "cause" and actual prejudice from the unconstitutional action.[32]

The Warren Court had also made it easier to obtain injunctions against improper state action. Where illegal raids had been made on a civil rights organization and the prosecutor had made public use of seized documents, the Court, in *Dombrowski v. Pfister,* allowed an injunction against prosecution under a broad statute infringing on free speech because First Amendment rights could not be sufficiently protected during a criminal trial. When lower federal courts interpreted this ruling broadly, there was a considerable increase in federal court challenges to state laws; these occurred even where state courts, given a chance, might have enforced federal constitutional rights.

Chief Justice Burger felt this practice overused federal courts and denigrated state court authority, and he was clearly determined to put an end to it. The Court's change of position came in 1971 in *Younger v. Harris.* The majority ruled in that case that federal courts were no longer to issue injunctions against state criminal prosecutions which had already been instituted, except where greater and immediate irreparable damage was threatened and that harm could not be prevented by raising the constitutional claims in a later state trial. This rule was to apply, said the Court, even when the challenged statute was clearly unconstitutional. However, in *Allee v. Medrano* the Court allowed an injunction against the Texas Rangers and a sheriff's department for harassing a farmworkers' organization through assaults, arrests, and jailings, because a continued pattern of activity was involved. Furthermore, where prosecution had not been initiated but was actually threatened, for example, where people handing out antiwar leaflets were told by police they would be (re)arrested if they continued their activity, the Court said that federal court declaration of one's rights was to be available even if one could not show the irreparable injury necessary for an injunction. (Whether declaratory relief could be used later to obtain an injunction divided the justices.)[33] The Court also increased rather than decreased access to the federal courts for relief from state laws by ruling that federal statutes did not prevent the granting of injunctions under the civil rights law (42 USC 1983) and that civil rights actions could be brought for violation of property rights—for example, in connection with wage garnishment—as well as of personal rights like free speech or the right to associate.[34]

After this apparent wavering, the doctrine of *Younger v. Harris* was extended. Thus the Court said a federal injunction could not be obtained against a state prosecution initiated after the federal challenge had been started or against civil proceedings closely related to criminal matters, for example, a state civil nuisance provision being applied to a theater showing allegedly obscene movies.[35] The former ruling led the dissenters to say that whether you could get an injunction depended on an improper "race to the courthouse" where the state could start later and get there first.

GOING UP: LAWYERS, GROUPS, AND APPEALS

Lawyers and Appeals

Once a case survives the hurdles just discussed, a court will decide the substantive issue posed by the case. Then consideration takes place among the lawyers as to whether the case should be taken further. Some formal rules affect what appeals may be brought—and when—but lawyers and supportive interest groups play a particularly important role in the decision to appeal a case either from the trial court to the courts of appeals or from the appeals courts to the Supreme Court. The result is that not all types of cases are appealed in the same proportions; moreover, in some cases, during an appeal attention may shift from the issues which were the focus of trial proceedings to other, more general legal issues.

Not all lawyers enter cases at the same stage. Over 85 percent of those who argued reapportionment and loyalty-security cases in the Supreme Court in the late 1950s and early 1960s were involved from the initial trial onward. By contrast, less than two-thirds of those who argued civil rights cases were involved in the initial trial; the remainder did not become involved until the Supreme Court stage. In criminal justice cases, where most attorneys became involved through court appointment, slightly over two-fifths were involved at the initial trial, another one-third-plus appeared in the case at the first appeal, and the remainder were first involved at the Supreme Court level. (Friendship was more likely than group affiliation to account for most lawyers' association with reapportionment and loyalty-security cases, while group affiliation explained most involvement in civil rights cases.)[36]

The importance of lawyers at all stages of litigation from initiation to final appeal cannot be underestimated, even though judges may have the final say in a case. Obviously lawyers help determine the case which will be brought initially and which form the universe from which appeals may be drawn. Decisions to initiate cases are often the result of implicit interaction between lawyers and judges. On the basis of the precedents established by judges' rulings, lawyers may try to discourage clients from pursuing litigation they are sure to lose. Yet where the law is not clear or fully developed, a lawyer wishing to solve a client's problem may bring to the courts questions the judges have not previously considered. Judges may encourage lawyers to bring cases by making hints or suggestions that particular legal issues have not been raised. On the other hand, by showing firmness in disposing of a matter, they will make clear that the court wishes to hear no more litigation on that subject.

The actions which the lawyers take at one stage of litigation affect—and may foreclose—later actions:

> In a court of first instance (trial court) the lawyer must be careful in shaping the record so that all relevant facts and issues are included. Possible grounds for appeal and constitutional issues must be established early in the process. The future course of a case in the appellate courts, or whether or not it ever reaches an appeal, depends in very large part on the way the case is presented in the trial court. The substance of a well-argued point and the well-reasoned brief is often represented in the court's opinion. This is particularly important if the issue goes beyond a technical point affecting only individual interests but rather involves questions the determination of which may have important ramifications for the entire political system.[37]

In our supposedly adversary system of justice, courts are expected to make their decisions on the basis of material submitted by the parties. This is particularly true with respect to factual matters at trial but is true as well, although less fully, at the appellate level on matters of law. Thus to a greater or lesser degree lawyers' positions place constraints on the decisions judges can reach. If both lawyers in a case press the same position or at least focus directly on the same narrow issue, the judges' freedom of action may be decreased. At other times, however, those constraints are not tight, particularly when lawyers plead multiple claims, press procedural and substantive questions simultaneously, or argue broad grounds about which they are personally indifferent. The justices also at times increase their freedom of action by developing additional information from their own research. This practice may be criticized because the litigants do not have a chance to comment on such judge-derived "inputs" for decisions, yet it may be necessary if courts and particularly the Supreme Court are to issue decisions affecting more than the immediate parties to the case.

By stimulating the development of a "new breed" of lawyers interested more in broader principles than individual litigation, the Supreme Court has helped itself produce the information on which broader-reaching rulings may be based. The Court's ruling in *Gideon v. Wainwright* requiring appointment of counsel for indigents in serious criminal trials, followed and reinforced by establishment of the "War on Poverty" Legal Services Program with its emphasis on "law reform" instead of individual "band-aid" law, helped produce lawyers broadly interested in the problems of the poor. Such lawyers were more likely than others to initiate broad legal challenges, for example, in the area of welfare policy. Their action reinforced the Warren Court's reach toward broad rules, which in turn further encouraged the lawyers.

As this discussion suggests, lawyers, whether they serve the government, private clients, or interest groups, do not have identical orientations to the law. Lawyers who have taken cases to the courts of appeals have been categorized into two groups. For one, law is basically used to resolve conflicts, with the lawyer acting as the client's agent. In the other, law—and a client's case—are seen more as a means for accomplishing social change; here the lawyer's view of the "social good" is particularly important in determining the course of litigation. The former view is more likely to be adopted by lawyers practicing corporate law, while the latter view is more congenial to those in other types of practice—criminal, environmental, consumer, labor, civil rights—and to women attorneys more than men.[38]

Appellate lawyers generally agreed on what factors should be considered in deciding whether or not to appeal and on the importance of those factors. The timing of a case, an organization's concern, and advice by other attorneys were among factors rated unimportant; chance of success was rated quite important. However, "social welfarists" were more likely than "entre-preneurials" to give the chance of winning greatest weight in deciding whether or not to take a case to the court of appeals and the "entrepreneuri-als" gave far less weight to "importance to society" than did the "social welfarists." A greater overall interest in obtaining a forum from which to make issues known led the latter to be more willing to file a petition for certiorari in the Supreme Court when the possibilities of its being granted were low. Women appellate attorneys appeared to need higher "odds" of winning than did the men before they would file a certiorari petition. They were also less likely than men to cite financial reasons, including a client's ability to pay, for appealing and placed more emphasis on strategic concerns than did men.

Other differences in lawyers' approaches to the law and to appellate litigation become clear from a look at lawyers who presented civil liberties cases to the Warren Court. Of particular note is that not all the lawyers who won landmark civil liberties rulings from the Court had set out to do so. Particularly in the area of criminal procedure, attorneys were often simply trying to win cases for their clients. In that effort, they argued constitutional questions along with everything else they could find to present, thus giving the Court its opportunity to establish the broad rules it announced. By comparison, attorneys who argued sit-in, reapportionment, and loyalty oath cases were far more likely to have had more in mind than to win the case for their client; they were interested in broader goals. In the sit-in cases, most lawyers wanted to overturn harsh punishments, but

> Others were primarily interested in using the litigation to outlaw discrimination generally. Some of the lawyers from the South hoped to use the litigation to teach their communities object lessons in tolerance, using the courts as a vehicle to impress upon their fellow citizens the illegitimacy and immorality of racial discrimination. Some wished to use the courts as a level to influence action by Congress, hoping to have the courts push Congress toward passage of civil rights legislation.[39]

Of particular significance in the Supreme Court is the participation of the solicitor general, who by representing the United States when it is a party and by appearing—more frequently since 1960—as a "friend of the Court" urges the executive branch's policies upon the Court. There is little question but that the Court pays close heed to his arguments. For example, in the sit-in cases, Solicitor General Archibald Cox, "trying to lead the Court down a path wide enough to protect the demonstrators and narrow enough to protect the Court as well,"[40] particularly from broad rulings which might later cause the Court problems, "consistently urged the Court to find in favor of the claims of the demonstrators, but usually on fairly narrow grounds"[41]—exactly the action the Court took.

The solicitor general's appearance in the Supreme Court is not an isolated act but is closely related to the Department of Justice's overall litigation posture. This includes the solicitor general's task of deciding what cases the government should appeal. When the government has lost in the lower courts, the solicitor general decides whether to argue the case for the agency, to let the agency "go it alone," or even to oppose the agency. Here he may "confess error," that is, tell the Supreme Court that the government should lose a case it has won in the lower courts—for example, because of improper actions like illegal wiretapping. Sometimes a decision not to "go forward" may be related to the Court's caseload;[42] at other times, such as "confession of error" situations, it may be based on need to protect the department's reputation and thus increase the chance of winning later cases. More often the decision is related to policy considerations which also affect the government's role concerning lower court litigation before it gets to the Supreme Court.

In trying to develop national policy, the government does not accept a single court of appeals ruling adverse to the government on a particular point of law as authoritative. Instead the government is willing to relitigate even within the same circuit if some basis can be found for distinguishing later cases from the initial ones. Only when three unanimous courts of appeals decisions have been decided against the government is the government willing to stop litigation on that point of law.[43] This continuous relitigation is part of an effort to create an intercircuit conflict which it is hoped the Supreme Court will accept, again showing the relationship between lower court and Supreme Court action. In one instance, the National Labor Relations Board lost in five different circuits in succession, and then won its point on the sixth try, creating the desired conflict.[44]

Despite the great weight the solicitor general carries with the Supreme Court, at least some justices will not automatically accept a confession of error. When the majority had followed a suggestion by the solicitor general that a case involving promises made to a government witness be returned to the trial court, Justices Rehnquist and Powell and Chief Justice Burger, dissenting, complained that "this Court does not, or at least should not, respond in Pavlovian fashion to confessions of error."[45] And several justices have also criticized the solicitor general for trying to involve the Court in solution of internal Justice Department problems. In 1975, in a case where criminal charges had been brought contrary to Department of Justice policy on "double jeopardy," the Court's majority remanded the case to allow the government to file a motion to dismiss the charges. The Chief Justice and Justices White and Rehnquist, all of whom had been Justice Department officials, objected, saying it was "not a judicial function and surely not the function of this Court" to help the government emphasize to its lawyers that policy should be consistently followed, particularly where a considerable amount of federal prosecutorial effort was going down the drain as a result. They felt it improper for federal courts always to "conform [their] judgments to results allegedly dictated by a policy, however wise, which the judicial branch had no part in formulating."[46]

Interest Group Involvement

Cases are usually brought in the name of individuals and are argued by individual lawyers. However, interest groups often assist with cases. Their involvement is essential because individuals generally cannot afford to finance the cost of a lawsuit. Including filing fees, purchase of a transcript, and the printing of briefs, such costs may exceed $20,000 just to bring a case to the certiorari petition stage even if lawyers have donated their time. (In the early 1950s, the five cases in *Brown v. Board of Education*, including the Supreme Court litigation, cost between $200,000 and $250,000; the figure would be much higher now because of inflation.) The principal reason for interest groups' involvement, particularly in "public interest" litigation and so-called "test cases" intended to test the validity of a statute or regulation, is the groups' interest in the policies affected by the cases. The participation of interest groups, even though it occurs in forms specific to the legal system, reinforces the judicial process's political character.

Interest group participation in cases is of several different types. Some groups have assisted with the expenses of a trial or an appeal. With causes célèbres like the Chicago, Harrisburg, or Wounded Knee political trials of the 1960s and early 1970s, "umbrella" groups may be established to assist in raising funds. Other groups become involved only at the appellate stage of a case as *amicus curiae* ("friend of the court") to argue legal points relevant to the case and of concern to the group's interests. *Amicus curiae* involvement is perhaps the most frequent litigation-related activity of interest groups.

"Friends of the court" have been allowed to present relevant argument because of the difficulties of intervening as a party in someone else's litigation. At first, friends of the court were thought to be neutral, that is, serving the court rather than the parties. However, this view was not reflected in groups' behavior, and overuse of *amicus* activity by the parties, who seemed to be engaged in a contest to "line up" the most "fans," led the Supreme Court to adopt rules which pose some obstacles to such participation. However, most groups still seek *amicus* participation to argue positions favorable to one of the parties. There is no right to participate in a case as *amicus curiae*. Under the Court's rules, if both parties do not consent, participation comes only on petition to the Court itself. However, the United States and its agencies (when sponsored by the solicitor general) or states or local subdivisions may participate without having to seek the Court's approval.[47] (Except for the United States government, those participating as *amicus* are seldom allowed to participate at oral argument.)

The role of *amicus curiae*, particularly that of the solicitor general, can be quite significant, particularly when *amicus*'s approach to a case differs from the principal parties' approach. For example, in *Mapp v. Ohio*, the parties argued the case on the issue of convicting someone for "mere possession" of obscene material, that is, without intent to sell, while *amicus* (the American Civil Liberties Union) focused on the admission of illegally seized evidence. The Court decided the second issue favorably to the ACLU in a landmark ruling excluding illegally seized evidence from state trials and did not reach the initial question.

More direct involvement occurs when the group provides the attorney from the beginning of a case. This gives the group a greater opportunity to shape the trial record instead of having to work on appeal with a record created by a lawyer without the group's interests as his or her central concern. While groups often become involved in cases when someone with a problem seeks out the group to ask for its assistance—and a commitment of its resources—the group may seek out litigants with a "case or controversy" in order to bring test cases. The individuals whose names are used must satisfy requirements such as standing, but it is really the group which is bringing the case. When the National Association for the Advancement of Colored People (NAACP) was accused of barratry—the stirring up of litigation in which it had no direct interest, usually known as "ambulance chasing"—for doing this, the Supreme Court said in *NAACP v. Button* that helping people protect their constitutional rights through litigation was protected as part of the First Amendment's right of association. At times, interest groups bring suits directly in their own names. This occurs where the group is trying to protect its own members, as the NAACP had to do when the South "counterattacked" after *Brown v. Board of Education* by requiring membership lists as a condition for "doing business," the subject of the protracted *NAACP v. Alabama* litigation,[48] and the just-noted barratry case. And increasingly, in part because of these rulings, it occurs in other areas of law as well, for example, in environmental litigation, often initiated by the Sierra Club or the Wilderness Society.

When an interest group supplies counsel and particularly when it seeks out litigants, a question may arise as to whether the case is the client's or the group's. Because a group's resources are limited, it is not likely to expend such resources on a case unless the group's own interests can be advanced. Those interests are sometimes defined by "group politics," that is, a group's relations with other groups, as when the NAACP's expansion into poverty litigation was a response to the activity of other civil rights groups in addition to being a reaction to developments in judicial doctrine.[49] However, the ethics of the legal profession require than an attorney act in the interest of the client. If the client shares the group's interests, there is no problem, but when interests are not identical, the lawyer may define the client's interests in terms of group interests rather than vice versa. For example, when the client is offered a settlement in a case which might "make new law" if the case were to proceed to trial, the lawyer might be tempted to recommend against acceptance of the settlement instead of "sacrificing" the group's investment in seeking new legal principle. Most groups say that the case is indeed the client's, not the group's, but instances can be found which indicate that the group has settled for a symbolic victory rather than more concrete rewards for the client or that the group has emphasized precedent-setting cases at the expense of actions to enforce the rights won in such cases.[50]

A group's association with a case or with a number of cases on a particular subject may leave the impression that the group had a well-developed litigation strategy. If we define strategy broadly as "overall plans, co-ordination and direction developed for a major area of litigation, general enough to allow for flexibility and adaptability to changing circumstances,"[51] that may be true in some situations. Some groups do define in advance the policy positions to which they will give preference in bringing or defending lawsuits for individuals who seek their aid and may go further in orchestrating an attack on a policy, as the NAACP did with respect to racial restrictive covenants on housing.[52] Yet one of two cases decided by the Supreme Court on this issue was not an "NAACP case" but had been filed by a lawyer not working with the group, an indication that groups are often propelled into a case before a strategy is developed. This also occurred with the sit-in cases, where no strategy existed because lawyers for the "Inc. Fund" (the Legal Defense and Educational Fund of the NAACP) initially had no idea that so many cases would develop.[53] What this suggests is that groups can have substantial influence over some cases and can participate in others but are not likely to be in complete control of very much of the litigation which is appealed to the Supreme Court.

Before a case can get to the Supreme Court, it has a "long way to go." As we have seen in this chapter, part of that "long way" involves the rules connected with the Constitution's "case or controversy" requirement—rules which, as interpreted by the Supreme Court itself, determine which cases may get into the courts in the first place. Some of those rules are relatively straightforward interpretations of "case or controversy," while others are based more on the justices' sense of what is appropriate, particularly in terms of the relationship between the national government and the states. Once these requirements are satisfied, lawyers' and groups' decisions as to whether to "go forward" (or "up") become crucial in the journey of a case to the Supreme Court. But the lawyers' and groups' decisions only provide the basic pool from which the Supreme Court takes its cases. Thus we must turn to look first at the cases actually brought to the Court, that is, at the size and content of its docket, and to the ways in which the justices "skim off" most of the cases so that they can have a relatively small number to which to give more extended attention.

NOTES

1. See Arthur Selwyn Miller and Jerome A. Barron, "The Supreme Court, the Adversary System, and the Flow of Information to the Justices: A Preliminary Inquiry," *Virginia Law Review* 61 (1975): 1187–1245; they cite libel and abortion as examples.
2. For a recent example of litigation "hoked up" so parties could get into federal court, see *Kramer v. Caribbean Mills*, 324 U.S. 823 (1969).
3. *National Railroad Passenger Corp. v. National Association of Railroad Passengers*, 414 U.S. 453 (1974); *Trbovich v. Mine Workers*, 404 U.S. 528 (1972); *Trafficante v. Metropolitan Life Insurance Co.*, 409 U.S. 205 (1972).
4. *Tileston v. Ullman*, 318 U.S. 44 (1943).
5. *Barrows v. Jackson*, 346 U.S. 259 (1953); *Eisenstadt v. Baird*, 405 U.S. 438 (1972); *Singleton v. Wulff*, 96 S.Ct. 2868 (1976).
6. *Peters v. Kiff*, 407 U.S. 493 (1972); *Taylor v. Louisiana*, 419 U.S. 522 (1972).
7. See *Data Processing Service v. Camp*, 397 U.S. 150 (1970); *Barlow v. Collins*, 397 U.S. 159 (1970).
8. *United States v. S.C.R.A.P.*, 412 U.S. 669 (1973).
9. *Frothingham v. Mellon*, 262 U.S. 447 (1923).
10. *United States v. Richardson*, 418 U.S. 166 (1974); *Schlesinger v. Reservists Committee to End the War*, 418 U.S. 208 (1974).
11. See also *Simon v. Eastern Kentucky Welfare Rights Organization*, 96 S. Ct. 1917 (1976), denying standing in a challenge to Internal Revenue Service rules said to encourage hospitals to limit their treatment of indigents. It would have to be shown that absent the rules the hospitals would provide the desired services.
12. *Hawaii v. Standard Oil of California*, 405 U.S. 251 (1972). The Court later prevented the state from suing on behalf of itself and local governments to recover overcharges resulting from antitrust violations, saying only "direct" purchasers—but not "indirect" ones—could bring such actions. *Illinois Brick Co. v. Illinois*, 97 S.Ct. 2061 (1977).
13. *Newman v. Piggie Park Enterprises*, 390 U.S. 400 (1968).
14. See *Moore v. Ogilvie*, 394 U.S. 814 (1969).
15. *Carafas v. Lavallee*, 391 U.S. 234 (1968).
16. *Griswold v. Connecticut*, 381 U.S. 479 (1965).
17. See *Noyd v. Bond*, 395 U.S. 683 (1969), and *Schlesinger v. Councilman*, 420 U.S. 738 (1975).

18. *Coleman v. Miller*, 307 U.S. 433 (1939); *Pacific States Telephone and Telegraph Co. v. Oregon*, 223 U.S. 118 (1912); *Luther v. Borden*, 7 How. 1 (1849); *Mississippi v. Johnson*, 4 Wall. 475 (1867).

19. *Gilligan v. Morgan*, 413 U.S. 1 (1973).

20. See Anthony D'Amato and Robert O'Neil, *The Judiciary and Vietnam* (New York: St. Martin's Press, 1972).

21. *First National City Bank v. Banco Nacional de Cuba*, 406 U.S. 759 (1972); *Alfred Dunhill of London v. Republic of Cuba*, 96 S. Ct. 1854 (1976).

22. *Monroe v. Pape*, 365 U.S. 167 (1962); *Moor v. Alameda County*, 411 U.S. 693 (1973).

23. *Georgia v. Rachel*, 383 U.S. 780 (1966) and *City of Greenwood v. Peacock*, 383 U.S. 808 (1966).

24. *Johnson v. Mississippi*, 421 U.S. 213 (1975).

25. *Lake Carriers Association v. MacMullan*, 406 U.S. 498 at 512 (1972). For a recent example, see *Bellotti v. Baird*, 96 S. Ct. 2857 (1976).

26. *Henry v. Mississippi*, 379 U.S. 443 (1965).

27. *Colorado River Water Conservation District v. United States*, 96 S.Ct. 1236 at 1244–45 (1976).

28. See *Cox Broadcasting Co. v. Cohn*, 420 U.S. 469 (1975).

29. *Hagans v. Lavine*, 415 U.S. 528 (1974); see also *Moor v. Alameda County*, 411 U.S. 693 (1973). In *Aldinger v. Howard*, 96 S.Ct. 2413 (1974), the Court held that pendent *parties* could not be joined to federal civil rights suits.

30. *Preiser v. Rodriguez*, 411 U.S. 475 (1973).

31. *Townsend v. Sain*, 372 U.S. 294 (1963); *Fay v. Noia*, 372 U.S. 391 (1963); *Picard v. Connor*, 404 U.S. 270 (1971).

32. *Brady v. United States*, 397 U.S. 742 (1970); *Parker v. North Carolina*, 397 U.S. 790 (1970); *Francis v. Henderson*, 96 S. Ct. 1708 (1976). The Court extended the doctrine of *Francis v. Henderson* in *Wainwright v. Sykes*, 97 S.Ct. 2497 (1977), to a defendant's failure to make timely objection to admission of his statements at trial. But see *Lefkowitz v. Newsome*, 420 U.S. 293 (1975), allowing some federal postconviction challenges despite guilty pleas when the states did so.

33. *Steffel v. Thompson*, 415 U.S. 452 (1974).

34. *Mitchum v. Foster*, 407 U.S. 225 (1972) and *Lynch v. Household Finance*, 405 U.S. 538 (1972).

35. *Hicks v. Miranda*, 422 U.S. 332 (1975); *Huffman v. Pursue*, 420 U.S. 592 (1975). See also the further extensions in *Juidice v. Vail*, 97 S.Ct. 1211 (1977) (state contempt actions) and *Trainor v. Hernandez*, 97 S.Ct. 1911 (1977) (state civil action for return of welfare payments bars federal court attack on state law concerning attachment of property).

36. Jonathan Casper, *Lawyers Before the Warren Court: Civil Liberties and Civil Rights, 1957–66* (Urbana: University of Illinois Press, 1972), tables 4 and 5, pp. 88–89.

37. Jeanne Hahn, "The NAACP Legal Defense and Educational Fund: Its Judicial Strategy and Tactics," in Stephen L. Wasby, *American Government and Politics* (New York: Scribner's, 1973), p. 396.

38. Gregory J. Rathjen, "Lawyers and the Appeals Process: An Analysis of the Appellate Lawyer's Beliefs, Attitudes, and Values," paper presented to the Midwest Political Science Association, 1975; Gregory J. Rathjen and Susan Ann

Kay, "The Impact of Sex Differences on the Attitudes and Beliefs of Appellate Lawyers," paper presented to the Southern Political Science Association, 1975. See also Rathjen, "Lawyers and the Appeals Process: A Profile," *Federal Bar Journal* 34 (Winter 1975): 21–41.

39. Casper, *Lawyers Before the Warren Court*, p. 145.
40. Joel B. Grossman, "A Model for Judicial Policy Analysis: The Supreme Court and the Sit-in Cases," *Frontiers of Judicial Research*, edited by Grossman and Joseph Tanenhaus (New York: John Wiley, 1969), p. 432.
41. Casper, *Lawyers Before the Warren Court*, p. 147.
42. Commission on Revision of the Federal Court Appellate System, *Structure and Internal Procedures: Recommendations* (Washington, D.C., 1975), p. 17.
43. Paul Carrington, "United States Appeals in Civil Cases: A Field and Statistical Study," *Houston Law Review* 11 (1974): 1101.
44. Commission on Revision of the Federal Court Appellate System, p. 59; see also appendix, pp. A-135ff.
45. *DeMarco v. United States*, 415 U.S. 449 (1974).
46. *Watts v. United States*, 422 U.S. 1032 at 1035–36 (1975).
47. For one discussion, see George C. Piper, "Amicus Curiae Participation—At the Court's Discretion," *Kentucky Law Journal* 55 (Summer 1967): 864–73. See also Samuel Krislow, "The Role of the Attorney General as Amicus Curiae," in Luther Huston et al., *Roles of the Attorney General of the United States* (Washington, D.C.: American Enterprise Institute, 1968), pp. 71–104.
48. See George R. Osborne, "The NAACP in Alabama," *The Third Branch of Government*, edited by C. Herman Pritchett and Alan F. Westin (New York: Harcourt, Brace, and World, 1963), pp. 234–74.
49. See Nathan Hakman, "The Supreme Court's Political Environment: The Processing of Noncommercial Litigation," *Frontiers of Judicial Research*, edited by Grossman and Tanenhaus, pp. 199–253, particularly p. 227.
50. See Stephen C. Halpern, "Assessing the Litigative Role of ACLU Chapters," *Civil Liberties: Policy and Policy Making*, edited by Stephen L. Wasby (Lexington, Mass.: Lexington Books, 1976), pp. 159–68, for a criticism of group strategy.
51. Hahn, "The NAACP Legal Defense and Educational Fund," p. 396.
52. Clement Vose, *Caucasians Only: The Supreme Court, the NAACP, and the Restrictive Covenant Cases* (Berkeley: University of California Press, 1959). An excellent recent study of interest group participation in litigation is Frank J. Sorauf, *The Wall of Separation: The Constitutional Politics of Church and State* (Princeton, N.J.: Princeton University Press, 1976), which discusses the American Jewish Congress, the American Civil Liberties Union, and Americans United [for Separation of Church and State].
53. Grossman, "A Model for Judicial Policy Analysis," p. 431, note.

6 THE SUPREME COURT: PRELIMINARY DECISIONS

THE COURT'S BUSINESS

Docket Size and Overload

The growth in the size of the Supreme Court's docket is affected both by factors external to the Court, such as the development of social problems and people's propensity to litigate, and by the Court's own actions. As noted earlier, the Court's consistent refusal to review cases on a particular subject will result in fewer cases of that type coming to the Court, while the Court's willingness to rule on a topic will increase the number being filed. Docket size also affects docket growth: a perception that the Court is overloaded may make people less inclined to ask for review.

The feeling that the Court's docket has grown had occurred not only in recent years but in earlier times as well. In the Civil War years (1862–1866) the Court handed down 240 decisions, but between 1886 and 1890 the figure had grown to 1,125. This paralleled increased district court workload—over 8,800 criminal cases in 1871 and 14,500 civil cases in 1873, but over 17,000 criminal cases disposed of and over 52,000 civil cases pending in 1900.[1] In 1870 the Supreme Court's docket contained 63 cases. Eight years later the number was 1,212. In 1891, the year of the Court of Appeals Act, the figure was 1,816, another 50 percent increase. This growth resulted largely from statutory changes, including provision for removal of cases to the federal courts and addition of federal question jurisdiction. Although the Court "found ways to voluntarily restrict jurisdiction," the Court's interpretation of the federal question jurisdiction added further to the docket. Despite justices' complaints that they were overworked, "a few important cases decided differently would have had the effect of discouraging much litigation."[2]

Despite earlier docket growth, fewer than 1,000 cases were filed in the Supreme Court annually as recently as the early 1940s. The number fluctuated between 1,000 and around 1,500 between 1944 and 1954, reaching a high of 1,510 in 1946. It then began a rather steady growth, passing 2,000 in 1961 and 3,000 only six years later. In 1972 over 4,500 cases were filed and the figure continued to climb to over 5,000.[3] In 1971 the docket stood at 4,515, three-and-a-half times the 1,353 cases of 1951. Roughly half of this increase was accounted for by petitions on the Miscellaneous Docket, principally prisoner petitions filed in forma pauperis, that is, by those too poor to provide the full set of printed materials. These petitions increased from almost 1,400 in the 1954 Term to over 3,800 in the 1973 Term.

In the 1963 Term the Court disposed of 2,401 cases—two on the Original Docket (original jurisdiction cases), 1,036 on the Appellate Docket ("paid" cases), and 1,363 on the Miscellaneous Docket. At the end of the year, 367 cases remained—mostly petitions for review not disposed of at term's end. The Court's output climbed continously after 1964 through the 1973 Term, when 3,876 cases were disposed of by the Court. End-of-term carry-over dropped from almost 600 in 1965 to 453 in 1966 but then grew, reaching almost 900 cases in the 1972 Term and just over 1,200 in the 1973 Term before falling to over 800 the next year.[4] These figures tell us that the Supreme Court's docket in the early 1970s was seven to eight times its size in 1925, the year the Court received its complete certiorari authority. The increased docket size did not lead, however, to comparable increases in the Court's decision of cases with full opinion. These remained at roughly a hundred per term into the late 1960s before growing in the 1970 Term to over 125 and to roughly 140 for the 1972 through 1975 Terms. In addition, each term the Court announced fifteen to twenty per curiam (unsigned) opinions announcing substantive law and substantially increased its use of summary judgments.

An indication of increased docket volume was the Court's action in October 1972 when it disposed of 708 cases in one day—mostly denials of review. The list for the first conference of the October 1975 Term contained 732 certiorari petitions and 61 appeals papers—867 cases—and on the first official day of the term, the Court disposed of 790 cases. In addition to noting probable jurisdiction in seven appeals and granting certiorari in nineteen cases, the justices affirmed twelve cases summarily; vacated eight others; dismissed thirty appeals; denied 619 certiorari petitions, nine habeas corpus petitions, eight mandamus petitions, and thirty-four petitions for rehearing; and handed down forty-five miscellaneous orders.

The size of the Court's docket has led some to say that the Court is "overloaded." This complaint had been made in earlier years even when dockets were much smaller, for example, in the nineteenth century before the courts of appeals were established and again before the certiorari jurisdiction was provided. However, it has been renewed with particular force in the 1970s, particularly by the Court's new members. The more senior justices were less likely to complain. In 1972 Justice Douglas, who said the courts of appeals were the overloaded courts whose judges had a much heavier opinion-writing load, called Supreme Court overwork a "myth" and stated that the Court was "if anything, underworked, not overworked." He pointed out that the Court had no backlog and was accepting roughly the same number of cases for argument as in the 1939 Term.[5] Justice Douglas was perhaps a "workhorse," saying that "no Justice of this court need work more than four days a week to carry his burden," which he claimed was "comfortable" even when he was hospitalized,[6] but other judges agreed with him. Justice Stewart, for example, joined Douglas in needling their colleagues for complaining about overwork when they insisted on taking cases of little importance.[7] However, a case that seems unimportant to some will seem quite important to others. For example, Justice White recently observed that "perhaps, in light of the current pressures on our docket, there may be a category of [intercircuit] conflicts . . . involving insignificant points of federal law, which we simply do not have the capacity to resolve." However, he disagreed with his colleagues that whether a private hospital funded by federal and state governments could refuse to perform abortions was such a case; instead he argued that the Court should resolve problems coming after its major decisions on abortion.[8]

White's comment suggests that it is not docket size per se which is the issue but the need for time to resolve certain types of cases. Another need is time for adequate deliberation. Justice Rehnquist has argued that for the Court to be current in its docket is insufficient: "It is essential that it have the necessary time for careful deliberation and reasoned decision of the very important types of cases which are the staple of its business today."[9] And Justice Blackmun has declared that "the heavier the burden, the less is the possibility of adequate performance and the greater is the probability of less-than-well considered adjudication," with the possibility of a "breaking point" beyond which a judge's work might become second-rate.[10] The justices have on occasion "expressly recognized time as a deterrent to adequate deliberation." This was perhaps most obvious in the *Reid v. Covert/ Kinsella v. Krueger* cases on courts-martial of wives of service personnel for murdering their husbands. After initially deciding the case one way (against the women), the Court, when it agreed to rehear the case, changed its mind as to the outcome and voided courts-martial of such civilians. The Court must allow itself adequate time for attention to those cases in which it wants to develop policy positions, so that "the contours of the change desired can be adopted and implemented in such a way that social stability rather than social upheaval is promoted."[11]

Chief Justice Burger has made several proposals to deal with the Court's docket. In addition to the pending proposal to establish a National Court of Appeals and the already-enacted elimination of most three-judge district courts so that cases would not come directly to the Supreme Court, Burger has also suggested that prospective appellants get approval from the courts before they could appeal their cases. The Commission on Revision of the Federal Court Appellate System, which also extensively examined the matter of overload, said that "overload" could not be evaluated in terms of cases filed. There was also a "hidden docket"—the cases in which the size of the Court's docket deterred people from asking for review because the probability of obtaining a decision on the merits was not worth the expense.[12] The commission stressed the need for adequate capacity for the declaration of national law and said that lack of such capacity led to lack of Supreme Court review for cases which "almost assuredly would have been taken twenty years ago."[13] This in turn led to unresolved conflict on points of law between the circuits and to delay as well as to uncertainty for both lawyers and lower court judges, who needed "a body of precedents adequate for confident decision."[14] Unresolved intercircuit conflict would also produce repetitive litigation, further increasing cases filed in the Supreme Court. Making the matter worse was that the Supreme Court would be forced to take cases "otherwise not worthy of its resources," because it was the only court capable of resolving conflict between the circuits.[15]

Docket Content

The size of the Court's docket is, of course, not all that matters; the content of that docket is of crucial importance. Over time the bulk of that content has shifted from nonconstitutional cases to ones which raise constitutional questions. As Justice Rehnquist pointed out, in 1973 over half the cases raised constitutional issues, with questions of federal statutory construction and intercourt conflicts accounting for most of the rest.[16] The growth in constitutional questions began in the 1930s. Constitutional law cases made up only 9 percent of the certiorari petitions filed in the 1929 and 1930 Terms but 14 percent in the 1937 and 1938 Terms. Included at the latter time were due process questions, which constituted the greatest portion, as well as issues based on the Commerce Clause, impairment of contract, and "full faith and credit." Cases based on the Bill of Rights were only 5 percent of the constitutional cases at that time. By the 1957 and 1958 Terms constitutional questions appeared in 37 percent of the cases filed; ten years later, the proportion was almost half (49 percent). It rose still further—to 58 percent—in the 1971 and 1972 Terms.

The proportion of cases entailing constitutional questions has always varied considerably between policy areas, approaching 100 percent through the entire recent period for state criminal cases. The increase in cases filed in that category helps explain the increase in the overall proportion of constitutional cases.[17] Criminal cases, roughly two-thirds of which were federal, made up 19 percent of the Court's appellate docket in the 1957 and 1958 Terms, but by the 1971 and 1972 Terms they constituted over one-third (35 percent) of the docket. Inclusion of the Miscellaneous Docket cases brings the proportion of criminal cases to over half—55 percent in 1957–1958, roughly 60 percent otherwise.[18]

From the late 1950s to the early 1970s, noticeable increases in cases filed occurred with respect to a number of policy areas. Practically the entire spectrum of federal criminal procedure matters showed increases, as did state cases on evidence, jury matters, right to counsel, search and seizure, and "procedure," as well as obscenity. "Growth areas" on the civil side of the Court's appellate docket included the military, racial discrimination, reapportionment, welfare, private antitrust, state government personnel, education, securities law, regulation of attorneys, and domestic relations. Over the same period there were decreases in cases on taxation, immigration, federal government personnel, federal contracts, the Interstate Commerce Act (private suits), patents-copyrights-trademarks, and the Federal Employers Liability Act; cases from some of the regulatory agencies (the FPC, FTC, and ICC) also decreased.[19]

If we turn from docket to disposition, we find that where prior to 1960 nonconstitutional holdings made up two-thirds to three-fourths of the Court's rulings, as a result of the Warren Court constitutional cases constituted one-half to two-thirds of the Court's full opinion decisions. As a result, "in each term the Supreme Court can be expected to hand down no more than 80, and perhaps as few as 55, plenary decisions in all areas of federal non-constitutional law."[20] During the Vinson Court and the Warren Court's first two terms, economic and civil liberties cases averaged 58 percent of all cases decided on the merits, but before the Burger Court produced changes in the contrary direction, the figure rose to 80 percent in the remaining Warren Court years—most of that increase coming from civil liberties cases.[21] The proportion of constitutional cases decided with full opinion did decrease in the 1973 and 1974 Terms, falling to about half in 1973 and to only 46 percent in 1974.[22] This was accompanied by an important shift in subject matter: "Dispositions with full opinions in economic matters rose from 39 percent in the 1968 Term to 55 percent in the 1972 Term, while the proportion of criminal and habeas corpus cases fell from 42 percent to 32 percent."[23] Yet trends are not always continuous; for example, in the 1976

Term, *three-fourths* of the Court's decisions dealt with civil liberties and closely related manners.

While private civil actions were increasing from 34 percent of the Court's docket in 1968 to 50 percent of the Court's docket in 1973–1974, there was a substantial drop in criminal cases from the courts of appeals to which the Supreme Court granted review. Thirty-five percent of the Court's full opinion cases in 1967 and 39 percent in the 1970 Term had involved criminal matters including habeas corpus, but for the 1973 and 1974 Terms, the figure had fallen to 27 percent. There were also changes between 1967 and 1973 in the sources of the cases the Supreme Court accepted for full opinion disposition. The number of appeals accepted in federal habeas corpus cases increased, while the number of federal and state criminal cases remained roughly the same. The lower federal courts supplied more civil cases, mostly involving state or local governments. During this period, with no clear trends, federal litigation ranged from one-third to 46 percent of the Court's full opinion cases; state and local litigation fluctuated from one-third up to half; and private litigation constituted the remainder, running from a low of 8 percent (1972) to over one-fifth (1967).[24]

Within the broad outlines just noted, the Court's full opinion cases have dealt with a wide range of topics. The range of cases with which the Court concerns itself is shown by a *partial* listing of issues decided during the October 1975 Term:

- discharge from job for pregnancy; disability pay for pregnancy
- permanent resident aliens' right to be federal employees
- state taxation of cigarettes on Indian reservation; state taxation of non-residents' income
- entrapment defense when government agent supplied drug
- political speeches and leafleting on military base
- right to counsel at summary court-martial
- Mine Safety Act procedures
- campaign finance legislation
- border patrol searches; seizure of personal business papers in home
- sympathy strikes in violation of union contracts; workers' refusal to work overtime
- "black lung" legislation
- parental and spousal consent requirements for abortions
- "check-off" of union dues for municipal employees' union
- illegitimates' rights under Social Security
- state aid to church-related colleges
- racial discrimination in admission to private school
- submission of zoning rules to referendum
- dismissal of striking teachers

- mandatory retirement at age fifty for state police
- right to hearing before discharge from job
- seniority rights after racial discrimination in employment
- libel of "public figures"
- "gag rules" on the press in reporting criminal trials
- jurisdiction of agencies to monitor radioactive wastes
- requiring federal installations discharging water and air pollutants to apply for state permits

Relatively few categories contain more than one or two cases, a result of the Court's deciding fewer than 150 cases and attempting to develop law on the many topics on which people have sought its guidance and direction. Recent exceptions have been few. In the 1973 Term, ten of ninety-three civil actions from the lower federal courts involved review of administrative action, six of those from the National Labor Relations Board; and there were five federal antitrust and three Freedom of Information Act (FOIA) cases in 1974, when four state and local cases involved welfare rights. Search and seizure questions have accounted for several cases in the federal criminal area in each of the last few terms, a reflection of the Burger Court's principal concern in the criminal justice field, and in the 1976 Term, which showed a continuation of that concern, there were *ten* cases on the question of "double jeopardy" alone.

By no means all the Court's rulings are in fields like racial discrimination, aid to parochial schools, free speech, and criminal procedure to which the media devote most attention—as one can see from the list above. Indeed, many recent cases have dealt with less "glamorous" subjects such as admiralty, bankruptcy, civil procedure, taxation, and patents. Such cases, however, may substantially affect the nation's economy; this is certainly true of rulings on labor-management relations and on complex mergers of economically distressed railroad corporations or Congress's establishment of Conrail. It is also true with respect to rulings concerning interstate commerce, which, while they do not bulk as large in the Court's docket as they did during the New Deal, still arise with regularity. An example was the Court's decision on the validity of a state corporation franchise tax; another, its invalidation as an interference with commerce of a state rule barring milk and milk products unless the state of origin accepts such products on a reciprocal basis.[25] Similarly, important questions of federalism must be decided—although they, too, are less visible than civil liberties cases. A recent example was the case testing whether state laws prohibiting employers from knowingly employing aliens not entitled to lawful residence in the United States were an unconsitutional regulation of immigration; the Court held in *DeCanas v. Bica* that they were not.

"SKIMMING OFF" I: DENYING REVIEW

The Court's decisions denying review and its other summary actions are of generally low visibility. Moreover, the reasons underlying them are often intentionally well hidden. The sheer numbers of such decisions and the variety of factors potentially playing a part in any particular Court action further mask their meaning but do not decrease their extreme importance, which stems from the fact that such decisions account for the bulk of the Court's actions, far exceeding in number the Court's formal statements of policy. For the Court to make a decision not to hear a case may be as important—not only for the litigants but also for the (unaware) public—as for it to decide particular controversial questions explicitly. The patterns of the Court's actions, reinforced by statements by some of the justices, provide strong evidence that the Court's actions denying review have clear policy implications.

The Supreme Court, as already noted, may pick and choose from among the cases people wish it to consider. The justices select for further consideration about half the appeals cases and a small and decreasing percentage of those brought up on certiorari. Although 17.5 percent of the certiorari cases were granted review in 1941, ten years later it was only somewhat more than 10 percent (11.1 percent), by 1961 it was 7.4 percent, and by 1971 it had reached 5.8 percent with a continued subsequent slow decrease.

In managing its docket, the Court must reduce the number of cases to a reasonable amount, but in doing so it does not pick cases at random or through some previously adopted formula. For one thing, there are some cases which the justices will find quite difficult to ignore. Included are challenges to major new federal statutes like the public accommodations section of the 1964 Civil Rights Act and to the 1965 Voting Rights Act and instances of extreme resistance to the Court's rulings, such as the Little Rock school desegregation situation when the Court not only departed from its pattern of not reviewing cases in that area but also held a special session to hear the case. These cases also show that other units of government—Congress, the president, and state officials—have definite effects on the Court's agenda. While there are some cases the Court cannot avoid, the justices must take care not to concentrate their efforts disproportionately in one or only a few areas of case law. Thus when it takes up one or more major cases on a particular subject, it is imposing limits on its ability to take more in the near future. Despite such limits, the Court does have considerable flexibility in case selection. The wide variety of issues which people want the justices to hear does allow the Court to avoid certain fields completely, at least in the short run, and its pattern of accepting cases for review varies from one policy area to another. In some areas the Court accepts virtually all cases, as it did with sit-in cases in the early 1960s, but in others it may take very few, as it did in school desegregation after *Brown v. Board of Education.*

In accepting cases for review, the Court must also be concerned with timing, which includes when the Court takes or rejects a case in relation to its own major doctrinal decisions. Accepting cases allows the Court to indicate that aspects of certain issues have not been settled or to prevent departure from its policy; rejecting cases allows the Court to show that certain questions are settled, that it does not wish to comment further on the issues. When the Court enters a policy area and then hesitates to follow through—true with school desegregation after *Brown*—disobedience to its will may be encouraged.

Docket management also involves selection of cases in terms of their breadth or narrowness. The Court may choose cases which focus directly on a broad issue, weeding out those with peripheral issues, or instead may select ones which are narrow, presenting only a limited aspect of a subject. Grouping cases on a topic provides the justices with a broader range of strategic alternatives and emphasizes the broad policy implications of the subject under review. Such choices are closely related to legal issues raised by lawyers. With a badly presented constitutional question, the Court can only accept a case knowing that the broad question will have to be decided. On the other hand, when lawyers have presented several arguments of varying breadth or argued several different grounds as underpinning for the result they wish to achieve, the Court's options are greater.

Such considerations take place in the context of the formal mechanisms by which the Court deals with the cases presented to it. Although the Supreme Court has come to have almost total control of what cases it hears, the formal rules are not without importance. The appeals jurisdiction is in legal theory mandatory; the Court is supposed to take and decide all appeals cases. They arise where:

- the highest state court invalidates a federal law or treaty as unconstitutional or upholds a state law or state constitutional provision against a challenge that it violates a federal law, a treaty, or the U.S. Constitution;
- a U.S. court of appeals declares a state law or constitutional provision unconstitutional or declares a federal law unconstitutional when the federal government is a party to a case;
- a federal district court declares a federal law unconstitutional, again where the United States is a party; or
- a three-judge district court has granted or denied an injunction in cases required to be brought before such a court.

Until recently, certain civil cases in federal district court in which the national government was a party—involving the antitrust, interstate commerce, and communications laws—also were in the appeals category, but much of this jurisdiction has been eliminated.

Despite the theoretically mandatory character of appeals, the Court has in effect made review of appeals discretionary. The requirement that appeals be decided leads the Court to give formal disposition to proportionately many more appeals cases than to those in the certiorari jurisdiction, but many appeals are disposed of summarily, a recognition of the Court's flexibility of action. As early as a hundred years ago, the Court allowed appeals (then writs of error) to be affirmed on motion without argument when the appeal appeared frivolous or appeared to have been undertaken for purposes of delay.

The important shift toward de facto discretionary treatment of appeals came in 1928, when the Court requested submission of a "jurisdictional statement" stating why the Court should take the case, later extended to indicating why the question was "substantial." The present situation is that for the Court to take an appeal from a state court, the case must contain a nonfrivolous federal question and be "substantial"; appeals from federal courts only require a showing of "substantiality." That a large percentage of appeals do not meet those standards is shown by the fact that over one-fourth of the appeals in the 1972–1973 Terms were "dismissed for want of a substantial federal question." Slightly under one-fifth were dismissed for want of jurisdiction, and under 5 percent were dismissed for miscellaneous reasons.[26] Dismissal for want of a substantial federal question may appear to be only a rejection of review. However, because of the idea that appeals must be decided by the Court, such a dismissal is formally considered to be a decision on the merits. (Dismissal for lack of jurisdiction is not a ruling on the merits.) As the Court recently reminded lawyers and judges about such cases:

> A federal constitutional issue was properly presented, it was within our appellate jurisdiction . . . and we had no discretion to refuse adjudication of the case on its merits. . . . We were not obligated to grant the case plenary consideration; and we did not; but we were required to deal with its merits.[27]

The Supreme Court expects lower courts to be bound by such rulings, just as it expects them to be bound by full opinion decisions. However, because the Supreme Court has not described the facts and issues covered or explained the basis for its dismissals, lower court judges must compare the cases they have under consideration with the Supreme Court's previous summary actions, a tedious task. Moreover, according to Justice Brennan, because the lower courts, being bound by the dismissals, should not look further at the issues posed in those cases, the Supreme Court is deprived of the lower courts' thinking on the complex problems often present in the cases. Brennan has also pointed out another confusing aspect of the Court's practice in disposing of appeals: federal appeals are affirmed while state appeals are dismissed for want of a substantial federal question. He suggests that the latter action could mean either that there was no substantial federal question or that one had been decided correctly below so that further consideration by the Supreme Court would not be fruitful.[28]

Cases are brought to the Supreme Court on certiorari, an order to a lower court to send up the records in a case, from the state courts when a decision on a federal question is favorable to the federal law, for example, when a state law is invalidated under the Supremacy Clause; from the courts of appeals in decisions involving interpretation or application of the Constitution, treaties, and federal laws; and in situations when those courts have ruled that state laws or constitutional provisions are *not* contrary to federal law. The Court received the first part of its certiorari authority in 1891 when the courts of appeals were established. Certiorari jurisdiction was extended in 1914–1916, when many state cases and some federal court of appeals cases were shifted to the Court's discretionary jurisdiction. The full growth of that jurisdiction did not come, however, until the Judges Bill of 1925.

The power to grant or deny certiorari is fully discretionary, and the Court can grant or deny it without giving any reasons. The justices may even change their minds once the writ has been granted, dismissing it as "improvidently granted," as they did in six cases from the 1970 and 1972 Terms. Such action may result when new developments have occurred after certiorari was granted or the justices discover more from briefs or oral argument than they knew from the certiorari petition. However, when this device was used to avoid decision of a cemetery discrimination case immediately after *Brown v. Board of Education,* it proved damaging to the Court's reputation because the Court had already heard oral argument in the case once and it dismissed the certiorari petition only after a request for rehearing had been made.[29]

Certiorari is by custom granted by a vote of at least four justices. Apparently the decision to grant review is seldom unanimous or nearly unanimous. Justice Brennan has pointed out that of the cases granted review in the 1972 Term, roughly three-fifths received only the four votes necessary or five votes, but in only 9 percent was the Court unanimous.[30] The judges voting against review are expected not to turn around and throw the case out but are to proceed to decide it on the merits. However, Justice Frankfurter at times refused to participate in deciding some cases when he thought certiorari should not have been granted; if all the justices had followed his practice, that would have turned the "rule of four" into a "rule of five." However, a 1976 dismissal of a certiorari grant as improvidently granted produced a complaint from Justice Brennan that the Court's standards for such action had been violated and that it was improper for a justice who had voted *not* to grant certiorari to turn around and vote to hold it improvidently granted unless those standards were met.[31] In some recent instances, for example, obscenity cases where the Court's alignments are firm, when four justices have voted to grant certiorari they have said they would not insist the case be heard because they knew that the other five would be in the majority on the merits.[32]

In the process by which the Court decides whether or not to grant certiorari, the Chief Justice plays a particularly important role because, assisted by his law clerks, he carries out the initial sifting of the certiorari petitions and also makes the conference presentation of most cases. (For the purpose of screening certiorari petitions, in a change initiated by Justice Powell, at least five or six justices now have "pooled" their clerks instead of having each justice's clerk go through the entire batch of petitions.) Until recently, the Chief Justice prepared a "dead list" of cases which were not discussed at conference unless such consideration was requested by another justice. This function was particularly significant for *in forma pauperis* petitions from prisoners—so-called "unpaid" petitions—of which there might be only one handwritten copy rather than one for each justice. All members of the Court now get copies of the papers on all the petitions, but the Chief Justice prepares a "discuss list." Cases not on that list are not discussed unless another justice specifically requests it; if such a case is an "unpaid" one, the requesting justice makes the conference presentation about it. Cases not on the "discuss list" and not added to it are automatically denied review. Once the Chief Justice has sorted out and "special listed" cases, review is granted to as many as one-third of the cases remaining.[33]

The granting of certiorari is discretionary but it is not random. Chief Justice Warren, in saying that "the standards by which the justices decide to grant or deny review are highly personalized and necessarily discretionary," claimed, "Those standards cannot be captured in any rule or guidelines that would be meaningful."[34] One need not, however, accept the last part of his statement fully. For one thing, talk of the Court's "informed discretion" or "informed arbitrariness" as key factors in granting review and Chief Justice Warren's comment that certiorari jurisdiction was "designed by Congress for a very special purpose ... not only to achieve control of its docket but also to establish our national priorities in constitutional and legal matters"[35] make clear that political—and strategic—considerations in the broadest sense are behind the Court's choices. Individual justices from time to time have indicated what the Court might be looking for. Chief Justice Vinson, in an off-the-court statement, once said that the Court does not grant certiorari merely to correct errors of the lower courts, but instead uses the writ to deal with cases with broader effects, "questions whose resolution will have immediate importance far beyond the particular facts and parties involved."[36] And Justice Harlan stated from the bench, "the certiorari jurisdiction was not conferred upon this Court 'merely to give the defeated party in the ... Court of Appeals another hearing,' ... or 'for the benefit of the particular litigants,' ... but to decide issues, 'the settlement of which is of importance to the public as distinguished from ... the parties.' "[37]

The Court has also set forth some reasons why it will grant certiorari. The Court's Rule 19, which stresses legal considerations, says that certiorari will be granted "only where there are special and important reasons therefore." Some—but not all—of the factors which the Court would *consider*, but which would not be determinative, are then indicated. These include, in addition to important questions the Supreme Court has not yet decided, intercircuit conflicts in legal interpretation, conflict between a lower court ruling and previous Supreme Court rulings, and lower court rulings which depart from "accepted canons of judicial proceedings."

The Court, however, does not often tell us much about why it has taken specific cases. In the 1956–1958 Terms, reasons for granting certiorari were given in about two-thirds of the Court's full opinion cases, but in 20 percent of the cases the reason was only "to decide the issues presented." Eliminating those, the reason was given in only 46.7 percent of the cases; the only often-mentioned reasons were a conflict between the circuits, the importance of the issue (mentioned more frequently), or both; why the case was important was not indicated.[38] In 200 cases from the 1970 and 1972 Terms, 88 contained clear reasons why the Court had granted review; most of the rest, particularly the Court's summary rulings, contained no reasons at all. In addition to miscellaneous reasons applicable to one or two cases such as the need to vacate an unclear ruling of a lower court or the granting of review in light of the "opinion of the Solicitor General," in fifteen cases conflict between lower courts was the reason cited; in eight, certiorari was granted specifically to determine whether one of the Court's previous decisions should be made retroactive; in seven, to deal with the applicability of recent decisions; and in five, to deal with "inconsistency" with or violation of previous rulings. Seven cases were said by the Court to involve constitutional matters such as due process and equal protection. In thirty cases, the Court said only that it was taking the cases to examine "important" or "substantial" or "novel" nonconstitutional federal questions.[39] That is not helpful when many "important" or "substantial" issues are denied review. In recent years these have included underrepresentation on a jury of blacks, women, and those in certain age brackets; discharge of a teacher for statements made in his home; double jeopardy claims; suspension of students for wearing "long hair"; allowing the Communist party on the election ballot; school busing; standards for appointment of counsel for indigent defendants; air pollution standards; and public housing.

Certain "cues" may help explain the Court's actions even if the justices will not provide explicit reasons for granting review. Two cues are the

THE SUPREME COURT: PRELIMINARY DECISIONS

source of the case and who is seeking review. In cases from the 1953–1956 Terms, there was little difference in the Court's treatment of state and federal certiorari petitions, although the latter were somewhat more likely to be granted. However, federal appeals were given much better treatment than state appeals; only 10 to 15 percent of the former were dismissed.[40] In 1947–1958, certiorari was granted at a rate of 49.1 percent when the federal government favored review and no other cue was available, but when all other parties sought review, it was granted only 5.8 percent of the time, the same rate as when no cue appeared. One study showed that when a civil liberties issue including race was the only cue present, it significantly (and positively) affected the granting of review, but presence of only an economic issue had little effect. Reanalysis, however, suggested little "cue" effect from civil liberties issues.[41]

Cue analysis has also shown that certiorari was granted in 12.8 percent of cases where the sole cues were dissension between lower courts or between judges in a single court. One study indicated that the Court "primarily heard cases in which the district and circuit courts agreed ... [and] hears many more unanimous appeals than split ones." Another researcher showed, however, that "the Justices were inclined to hear reversed decisions more than affirmed ones, non-unanimous decisions" more than unanimous decisions, and en banc decisions more than panel decisions" from the courts of appeals, with disagreement within a court of appeals providing a stronger signal to the justices than did disagreement between levels. The conclusion of the first study was that, overall, "the Supreme Court ... is not primarily guided by lower court agreement—either intercourt or intracourt—in deciding the cases it will hear or reverse."[42]

If cues present in certiorari petitions may help us understand the nonrandom nature of the Court's exercise of discretion in granting review, arguments about the meaning of the Court's denial of review may also help. The Court gives reasons here even less frequently than when it grants review. A sample of more than 3,000 denied petitions over twenty years produced less than forty explanations, of which the most common was dismissal on motion of the parties or failure to timely file.[43] In the 1972 and 1973 Terms, 60 percent of the dissenting opinions in certiorari denials stated the need for a national decision to resolve conflicts between lower courts or between the lower courts and the Supreme Court, the presence of statutory questions which needed to be resolved, or "the existence of important questions for decision."[44]

The formal meaning of certiorari denial is only that the Court has not decided the case. Such a position was held even by Justice Douglas, an extremely frequent dissenter from certiorari denials: "Our denial of certiorari imparts no implication or inference concerning the Court's view of the merits. . . ."[45] However, some observers of the Court do not accept this statement. Fuel is added to the fire when the justices cite the Court's certiorari denials in concurring and dissenting opinions, at times more than as a matter of form. Recognizing the effect of certiorari denials, Chief Justice Warren wrote after his retirement, "Denials can and do have a significant impact on the ordering of constitutional and legal priorities. Many potential and important developments in the law have been frustrated, at least temporarily, by a denial of certiorari."[46] While he stopped short of saying such action was purposeful, disagreement within the Court indicates less than full acceptance of the Douglas-enunciated position. Justice Blackmun speculated in 1973 that dissents from an earlier certiorari denial were "not without some significance as to [the justices'] and the Court's attitude." This provoked an extended response from Justice Marshall, who said justices may simultaneously agree about an issue's importance and feel the case was not the "appropriate vehicle for determination of that issue." Perhaps too defensively, Marshall even attacked speculation about reasons why certiorari is denied. He said that "the point of our use of a discretionary writ is precisely to prohibit that kind of speculation" and asserted, "Reliance on denial of certiorari for *any* proposition impairs the vitality of the discretion we exercise in controlling the cases we hear."[47]

Outside the Court, "the inferences that people will draw from actions taken by the Supreme Court cannot be controlled and those who are result-oriented may read consideration of the merits" into denials of review.[48] As Adamany remarked, "When the Court consistently leaves undisturbed decisions at variance with principle, or when it denies certiorari in a notorious case ... the public may well believe that the Court is implementing an unspoken constitutional judgment."[49] Such inference drawing is further stimulated by lawyers' continued citation of certiorari denials in the lower courts. Although such citation may mean only that the lawyer feels that the matter is not settled and should therefore be decided, that is a stronger stance than a position that "the Court didn't hear it."

Dissents from certiorari denials do indicate that such denial entails implicit policy making. There has been a recent substantial increase in such dissent —a more than threefold increase over a period of four terms of Court— despite a norm against such votes.[50] Much of that increase was accounted for by Justice Douglas, who dissented alone from 169 certiorari denials in the 1970 Term (he wrote opinions in only five of these) and from 272 certiorari denials in the 1972 Term. However, there were 81 other dissents in the 1970 Term and 35 others in the 1972 Term, with 24 opinions accompanying the dissents in the latter term.[51] The Hruska Commission, pointing out that at least one justice not only noted his dissent but also wrote an opinion in more than eighty certiorari denials in the 1972 and 1973 Terms combined, said this number understated the amount of disagreement both because of the norm against voicing dissents but also because a justice might not so vote unless he could write an opinion—for which he might well not have the time.

Such dissents are quite likely to come from justices of similar ideological positions, such as the liberals Douglas, Brennan, and Marshall in the early years of the Burger Court. For example, where the government had won a criminal case in the lower courts, a justice who generally favors a criminal defendant (as shown by his votes in full opinion cases) would vote to grant review, while justices who would support the government once the case was accepted would vote not to grant certiorari. And a judge's certiorari votes did predict his votes on the merits in cases the Court accepted in the 1947–1956 Terms. Liberal and conservative voting blocs among the justices could be identified not only from votes on the merits but also at the certiorari stage: a justice who voted with one set of justices in full opinion cases was *un*likely to vote with his remaining colleagues at the review-granting stage.[52] It has been suggested that justices opposed to the trend of the Court's decisions might vote against granting certiorari in order to avoid having the Court make more "bad law," but in recent years this appears not to have happened; in any event, the suggestion still is based on the idea that justices of similar ideology will vote together in granting or denying review.

The most obvious indication that the justices engage in strategy with respect to granting review has been shown with respect to an earlier period. Schubert found that during 1942–1948 in Federal Employees' Liability Act cases, a bloc of four justices (Murphy, Rutledge, Black, and Douglas) seemed to vote together on certiorari so as to achieve victory on the merits. Their voting paralleled a strategy of never voting for the railroad (the employer) but always voting for the worker on certiorari when the court of appeals had reversed a proworker district court decision, then always voting for the worker on the merits. The four had a 92 percent success rate (twelve of thirteen) when they followed that pattern. When they voted for certiorari when workers had lost in both the trial and appeals courts, they had a 73 percent success rate (eight of thirteen). In the 1956–1957 Terms, when the four-judge bloc grew to five, they "won" thirteen of fourteen cases, with the "lost" case one in which no justice voted for certiorari because the appeal was so frivolous.[53]

Further confirmation of the policy implications of certiorari comes from the Court's disposition of cases to which it does grant review. The Court's reversal of the lower courts in roughly two out of three certiorari cases in the 1949–1951 Terms and its similar frequent reversal of the lower courts in such cases in the 1953–1956 Terms provided a pattern "too definite to have arisen by sheer happenstance,"[54] and seemed to indicate a better-than-even chance that the Court approved of lower court decisions to which it did not grant review. That certiorari is granted primarily so that lower court decisions can be reversed and so that the Supreme Court can state its own values and develop its own policy has been confirmed in other studies. In the 1967–1972 period, the Supreme Court reversed the federal courts at a 82.7 percent rate and the state courts in 91.1 percent of the cases. The figures were somewhat lower for the Court's formal opinions (slightly over 70 percent for both federal and state courts) as a result of a "fall-off" in reversal from the 1967–1968 Terms to the 1969–1972 Terms. They were, however, much higher in *per curiam* decisions in which substantive law was announced—97.4 percent for the state courts, 88.1 percent for the federal courts. In *per curiams* not announcing law and in the Court's other summary orders, virtually all lower court rulings were reversed—in 96.1 percent of the federal and 98.8 percent of the state cases.[55]

Reversals of lower court rulings are also quite likely in cases in the Court's appeals jurisdiction, particularly in those disposed of summarily. In 1953–1956, appeals from state courts were more likely to be reversed than were those from federal courts, which were affirmed more frequently than they were reversed. However, federal court appeals were particularly likely to be reversed in formal opinions—so that the Court could explain the law to the lower federal courts. In the 1967–1972 Terms, somewhat over three-fifths of federal appeals were affirmed, while over three-fourths of state appeals were reversed. In appeals decided with formal opinions, both federal and state appeals were reversed more than three times out of four. *Per curiam* appeals decisions announcing substantive law, although few in number, produced more reversals than affirmances. In nonsubstantive *per curiam* rulings and orders, federal appeals were affirmed at a high rate (77.1 percent, but over 85 percent for 1967–1968), while there was an extremely high rate of reversals (92.3 percent) in state cases.

Giving the Court the power to decide which cases to accept has produced changes in the Court's decision-making "style." When the Court had to decide all cases brought to it but its caseload was still small, the justices would decide each case carefully and deliberately—that is, they could engage in rational decision making. With time to consider each case, the likelihood that they would depart from precedent would be greater than otherwise.[56] With an increased caseload but still operating under the pre-1925 requirement of hearing all cases, decision making was incremental. Changes from past cases were limited, with precedent serving as a cue to allow prompt disposition. Instead of making law, the Court was more likely to be engaged in interpretation of existing doctrine. With the power to turn away cases, the Court could return to rational decision making, at least for those cases it accepted, thus allowing more policy innovation. With the caseload again growing, "mixed-scanning" decision making is produced. A quick glance is given to most cases before they are rejected, while those accepted get more extended and careful treatment. Here the justices' basic doctrinal attitudes affect not only the Court's full decisions but its decisions to grant review as well. The Court's ability to innovate may, however, be hindered if docket size gets too large and approaches "overload chaos," an unmanageable situation which some justices suggest now exists.

Changes besides decision-making style accompany the shift from mandatory to discretionary review, such as an increase in dissenting votes and dissenting opinions. After 1925, this first appeared in the appeals jurisdiction, from which less important cases had been removed and shifted to the certiorari category. The more controversial appeals cases, which would naturally produce more dissent, were the ones which remained. Dissent later increased—as did the number of concurring opinions—in certiorari cases as well, as winnowing progressed there.[57]

"SKIMMING OFF" II: SUMMARY DISPOSITIONS

Differences between results in the Court's full opinion cases and those decided summarily requires some attention to the latter, which are often ignored despite the important policy issues given such treatment. During the 1975 Term, state fair campaign practices laws, spanking of pupils, penalties on auto manufacturers for refusing to recall defective cars, and laws concerning "crimes against nature" (applicable to homosexuals) were all handled through summary disposition. Although critics have suggested that the disposition of a case often receives insufficient attention,[58] the decision to grant review in a case and the disposition of a case are linked. Summary disposition provides the justices an additional option when they wish to take actions on cases for which the Court may feel it does not have time for full-dress treatment including oral argument.

Summary dispositions include a variety of brief orders and some *per curiam* (unsigned) rulings, although other *per curiams* are indistinguishable from full signed opinions in announcing substantive law except perhaps for their relative brevity. First used to indicate only cases where the substantive law was "indisputably clear," *per curiam* rulings came to cover orders in original jurisdiction cases, dismissals of appeals for want of a substantial federal question, and obviously moot cases. Now, however, most cases not announcing new substantive law are placed with the Court's other orders (such as those granting and denying review), where they appear in such new categories as "affirmed on appeal" and "vacated and remanded on appeal."

Use of a *per curiam* ruling rather than a signed opinion may indicate that the case is considered routine; the Court may also wish to signal that the outcome is obvious and should receive prompt compliance—as in the 1969 ruling in *Alexander v. Holmes County* ending "all deliberate speed" in school desegregation. Another use of the *per curiam* opinion comes when the Court is very badly divided; the opinion is limited to the policy on which the justices can agree, while each spells out his reasoning separately. Examples are provided by *Furman v. Georgia,* the 1972 death penalty cases where each majority justice appended a separate opinion to the brief *per curiam* ruling, and *Buckley v. Valeo,* the 1976 ruling on campaign finances where five justices dissented from various parts of the ruling so that a single opinion of the Court could not be easily identified.

Per curiam rulings have been criticized on several grounds:

> (1) In some instances the Court appears to use per curiams to avoid an expression of opinion on important issues. (2) Frequently the Court does not make clear what is decided. (3) At times the law appears to be altered with little if any expressed explanation or justification.[59]

Similar criticisms have been made of summary judgments without opinion because a lower court judge or lawyers receives little guidance from the citation of one or two cases accompanying an order. One cannot tell whether the Court has merely affirmed the lower court's result or adopted its reasoning, true even when the Court cites one of its own earlier rulings for support. For example, Justice Brennan argued that the Court's approach to obscenity cases prior to 1973 hopelessly confused the lower courts because the justices reversed convictions while writing no opinion and citing only *Redrup v. New York,* a case in which the Court conceded its internal disagreement about obscenity. Such action by the Court settled particular cases "but offer[ed] only the most obscure guidance to legislation, adjudication by other courts, and primary conduct," with the Court "deliberately and effectively obscur[ing] the rationale underlying the decision."[60] And the Hruska Commission has said that summary affirmances or dismissals of appeals "cannot be considered the equivalent of plenary disposition for purposes of providing an adequate body of precedents on recurring issues of national law."[61]

The Court has now compounded these difficulties by indicating that summary rulings are not to be given as much precedential weight as a fully developed Court opinion.[62] According to Chief Justice Burger, not only does a summary affirmance not necessarily mean that the Supreme Court is adopting the lower court's ruling, but the Court need not adhere to a position appearing to result from several summary actions. Thus "the Court has not hesitated to discard a rule which a line of summary affirmances may appear to have established" when the justices later give fuller attention to another case raising the same issue.[63]

Despite this criticism, summary dispositions allow the Court to "clean up" a large number of cases in a particular policy area. They may also be used to make clear to lower court judges that their rulings need to be reexamined and cases disposed of more completely in the lower courts. Summary actions

may also preserve freedom of action for the justices, by providing a result
and thus perhaps an implicit message, without the constraints which full
development of doctrinal reasons would impose. However, a summary action's visibility may be too low for the ruling to have its intended effect,
particularly when resistance to the Court's policy exists. Thus the Court's
mid-1960s attempt to communicate through *per curiam* rulings that desegregation should take place more rapidly was generally ignored, requiring
more explicit rulings in 1968 and 1969.

The growth in the Supreme Court's docket and changes in the content
of that docket have now been examined, along with the quesiton of whether
the Court is "overloaded." The justices' exercise of discretion in its certiorari
jurisdiction has also been discussed and the possible meanings of the Court's
denial of review explored. The Court's use of its certiorari jurisdiction and
its conversion of the theoretically mandatory appeals jurisdiction into discretionary jurisdiction are ways in which the Court "skims off" cases. Another way, also noted here, is the use of summary dispositions on the merits.
These actions, important aspects of the Court's strategy, help the justices
pick the "right case at the right time" for more intensive examination. In the
next chapter, we look at the procedures by which such cases are handled,
the patterns of the justices' votes, and the writing and release of the Court's
opinions.

NOTES

1. Lawrence Friedman, *A History of American Law* (New York: Simon & Schuster, 1973), pp. 332, 335.
2. Mary Cornelia Porter, "Politics, Ideology and the Workload of the Supreme Court: Some Historical Perspectives," paper presented to the Midwest Political Science Association, 1975, pp. 8, 10.
3. Figures from Gerhard Casper and Richard Posner, "A Study of the Supreme Court's Caseload," *Journal of Legal Studies* 3 (1974): 340 (table 1).
4. Figures from Statistical Tables in the annual November issue of *Harvard Law Review* on the Supreme Court's term, particularly Five Year Table I, 62 (November 1968): 310 and 87 (November 1973): 310–11, and Table II, 88 (November 1974): 277 and 89 (November 1975): 278.
5. *Tidewater Oil Co. v. United States,* 409 U.S. 151 at 174–76 (1972).
6. *Warth v. Seldin,* 422 U.S. 490 at 519 (1975).
7. See *Butz v. Glover Livestock Commission Corp.,* 411 U.S. 182 at 189 (1973).
8. *Greco v. Orange Memorial Hospital Corp.,* 96 S.Ct. 433 at 436 (1975). See also *Bailey v. Weinberger,* cert. denied, 419 U.S. 953 (1974).
9. William H. Rehnquist, "The Supreme Court: Past and Present," *American Bar Association Journal* 59 (April 1973): 363.

10. Justice Blackmun to Senator Hruska, in Commission on Revision of Federal Court Appellate System, *Structure and Internal Procedures* (Washington, D.C., 1975), pp. A-237–A-238.

11. S. Sidney Ulmer and John Alan Stookey, "How Is the Ox Being Gored: Toward a Theory of Docket Size and Innovation in the U.S. Supreme Court," *University of Toledo Law Review* 7 (Fall 1975): 7, 9–10.

12. Commission on Revision of the Federal Court Appellate System, *Structure and Internal Procedures*, p. 17.

13. Ibid., Justice Blackmun to Senator Hruska, April 30, 1975, p. A-238.

14. Ibid., p. 29.

15. Ibid., p. 31.

16. Rehnquist, "The Supreme Court," p. 363.

17. Casper and Posner, "A Study of the Supreme Court's Caseload," table 3, p. 343, and table 6, p. 350.

18. Ibid., table 8, p. 351, and table 9, p. 355.

19. Ibid., tables 11–13, pp. 356–58.

20. Commission on Revision of Federal Court Appellate System, p. 19.

21. Glendon Schubert, *The Constitutional Polity* (Boston: Boston University Press, 1970), p. 10.

22. Statistics derived from *Harvard Law Review* tables.

23. J. Woodford Howard, Jr., "Is the Burger Court a Nixon Court?" *Emory Law Journal* 23 (Summer 1974): 757.

24. Data from *Harvard Law Review* tables.

25. *Colonial Pipeline v. Traigle*, 421 U.S. 100 (1975); *Great Atlantic and Pacific Tea Co. v. Cottrell*, 96 S.Ct. 923 (1976).

26. See David Rohde and Harold Spaeth, *Supreme Court Decision-Making* (San Francisco: W. H. Freeman, 1975), table 6.1, p. 120.

27. *Hicks v. Miranda*, 422 U.S. 332 at 344 (1975), also quoting Justice Brennan, *Ohio ex rel. Eaton v. Price*, 360 U.S. 246 at 247 (1959), to the same effect.

28. *Colorado Springs Amusements v. Rizzo*, 96 S. Ct. 3228 (1976). See also Justice Brennan's dissent in *Sidle v. Majors*, 97 S.Ct. 366 (1976).

29. *Rice v. Sioux City Memorial Park Cemetery*, 349 U.S. 70 (1955).

30. William Brennan, "The National Court of Appeals: Another Dissent," *University of Chicago Law Review* 40 (1973): 479.

31. See *Burrell v. McCray*, 96 S.Ct. 2640 (1976).

32. See, for example, *Dyke v. Georgia*, 421 U.S. 952 (1975). Justice Stevens has now suggested that such an alignment is a sufficient reason for them now to vote to deny certiorari. *Liles v. Oregon*, 96 S.Ct. 1749 (1976).

33. S. Sidney Ulmer, William Hintze, and Louise Kirklosky, " The Decision to Grant or Deny Certiorari: Further Consideration of Cue Theory," *Law & Society Review* 6 (May 1972): 640.

34. "Retired Chief Justice Warren Attacks ... Freund Study Group's Composition and Proposal," *American Bar Association Journal* 59 (July 1973): 728. This section is a revised version of Stephen L. Wasby, "The Supreme Court's Docket, Discretionary Jurisdiction, and Judicial Behavior Studies," paper presented to the Midwest Political Science Association, 1975, and revised for Wasby, *Continuity*

and Change: From the Warren Court to the Burger Court (Pacific Palisades, Calif.: Goodyear Publishing, 1976), pp. 33–40.

35. "Retired Chief Justice Warren Attacks ...," p. 728.
36. Fred M. Vinson, "The Work of the Federal Courts," *Courts, Judges, and Politics,* edited by Walter F. Murphy and C. Herman Pritchett (New York: Random House, 1961), pp. 55–56.
37. 392 U.S. at 478–79.
38. Joseph Tanenhaus, Marvin Schick, Matthew Muraskin, and Daniel Rosen, "The Supreme Court's Jurisdiction: Cue Theory," *Judicial Decision-Making,* edited by Glendon Schubert (New York: Free Press, 1963), p. 114.
39. Data collected and analyzed by John Rink, Southern Illinois University at Carbondale.
40. Glendon Schubert, *Quantitative Analysis of Judicial Behavior* (Glencoe, Ill.: Free Press, 1959), pp. 53–55.
41. Tanenhaus et al., "The Supreme Court's Jurisdiction," p. 123; S. Sidney Ulmer, "Revising the Jurisdiction of the Supreme Court," *Minnesota Law Review* 58 (November 1973): 154; Ulmer et al., "The Decision to Grant or Deny Certiorari," pp. 640–41.
42. Richard J. Richardson and Kenneth N. Vines, *The Politics of Federal Courts* (Boston: Little, Brown, 1970), p. 153; J. Woodford Howard, "Litigation Flow in Three United States Courts of Appeals," *Law & Society Review* 8 (Fall 1973): 47; Richardson and Vines, p. 155.
43. Tanenhaus et al., "The Supreme Court's Jurisdiction," p. 114. But see *Chapman v. California,* 405 U.S. 1020 (1972), where lack of finality of a state ruling is given as a reason.
44. Commission on Revision of the Federal Court Appellate System, *Structure and Internal Procedures,* pp. 48–49.
45. *Farr v. Pitchess,* 409 U.S. 1243 at 1246 (1973), citing *Maryland v. Baltimore Radio Show,* 338 U.S. 912 at 919 (1950).
46. "Retired Chief Justice Warren Attacks ...," p. 728.
47. *United States v. Kras,* 409 U.S. 434 at 443 (Blackmun), 461 (Marshall) (1973).
48. Thomas Lewis, "The Sit-in Cases: Great Expectations," *Supreme Court Review* 1963, edited by Phillip Kurland (Chicago: University of Chicago Press, 1963), p. 132, note.
49. David Adamany, "Legitimacy, Realigning Elections, and the Supreme Court," *Wisconsin Law Review* (1973), p. 801, drawing on Jan Deutsch, "Neutrality, Legitimacy, and the Supreme Court," *Stanford Law Review* 20 (1968): 207.
50. Commission on Revision of the Federal Court Appellate System, *Structure and Internal Procedures,* p. 48.
51. Data gathered by John Rink, Southern Illinois University at Carbondale.
52. See S. Sidney Ulmer, "Supreme Court Justices as Strict and Not-so-Strict Constructionists: Some Implications," *Law and Society Review* 8 (Fall 1973): 27–28, and Ulmer, "Voting Blocs and 'Access' to the Supreme Court: 1947–1956 Terms," *Jurimetrics Journal* 16 (Fall 1965): 8, 12.
53. Glendon Schubert, "The Certiorari Game," *Quantitative Analysis,* pp. 210–54.
54. Schubert, *Quantitative Analysis,* p. 66. For his data, see pp. 55–67.

55. Analysis of data compiled by David Gruenenfelder, Southern Illinois University at Carbondale. See also Howard, "Litigation Flow," p. 48.
56. The argument is that of Ulmer and Stookey, "How Is the Ox Being Gored"; see also John A. Stookey, "Possible Linkages Between Jurisdictional Change and Policy Output in the Supreme Court," paper presented to the Political Science Association, 1975, which contains some discussion not included in the Ulmer-Stookey article.
57. Stephen C. Halpern and Kenneth N. Vines, "Institutional Disunity, The Judges' Bill and the Role of the U.S. Supreme Court," *Western Political Quarterly*, (Fall 1977).
58. See Note, "Individualized Criminal Justice in the Supreme Court: A Study of Dispositional Decision Making," *Harvard Law Review* 81 (April 1968): 1260–79.
59. Note, "Per Curiam Decisions of the Supreme Court: 1957 Term," *University of Chicago Law Review* 26 (Winter 1959): 282.
60. *Paris Adult Theater I v. Slaton*, 413 U.S. 49 at 83 (1973).
61. Commission on Revision of the Federal Court Appellate System, p. 14.
62. See *Edelman v. Jordan*, 415 U.S. 651 at 671 (1974) (Justice Rehnquist).
63. *Fusari v. Steinberg*, 419 U.S. 379 at 391–93 (1975).

7 | THE SUPREME COURT: FULL-DRESS TREATMENT

PROCEDURE IN THE COURT

The Supreme Court's annual term begins in October and has usually ended by late June. However, in 1974 the Nixon tapes case helped carry the Court into July and in 1976 the term also continued past July 1, and the Court has also held special sessions in 1958 for the Little Rock school desegregation case and in 1972 for the Democratic National Convention delegate challenges. After lawyers have filed briefs subsequent to the granting of review, the cases in which the justices want to hear oral argument are argued before the Court from the beginning of the term through late March or early April. Cases to which the Court grants review early in the term may well be argued and decided in the same term; those accepted later in the term are not likely to be argued until the following term. The Court hands down decisions in all the term's argued cases before the term ends unless a case is set for reargument; this would result if the Court were badly split or if the justices wanted to wait until a vacancy were filled before it made a decision.

The justices meet in conference throughout the term, with the exception of Christmas and Easter recesses and recesses for opinion writing; they now meet twice a week—on Wednesday and Friday—during weeks when cases are being argued. In the conference, the Chief Justice makes the initial presentation of a case, including his own comments on it. Each justice, the most senior justice first and the most junior last, then comments in turn. Although discussion often indicates justices' votes, a formal vote is taken, in order of reverse seniority, the most junior justice first so that he would not be influenced by his seniors' votes. The initial conference vote is considered only tentative, so justices may change positions before the final vote is taken later. Thus what began as the majority at times becomes the minority.

After the initial vote, the Chief Justice assigns the task of writing the Court's opinion; if he is not in the majority, the most senior majority justice makes the assignment. The justice writing the Court's opinion circulates a draft to all the other justices for comments. After receiving his colleagues' responses, he may make changes; if he does, he will circulate the revised draft. Each justice must have the opportunity to see the opinions that every other justice has written so that he may express additional views in a concurring opinion or write an opinion dissenting from the majority's. (No one assigns the writing of a minority opinion, the dissenters deciding among themselves who will write.) If some majority justices join the Court's judgment, its decision or result, but do not join in its opinion, there may be only a "plurality" or "prevailing" opinion. This occurs when as many (or more) justices dissent as support the basic opinion, for example a 5-to-4 vote with the five majority justices themselves split 3-to-2 over the proper opinion. A plurality opinion has much less precedential value than an opinion of the Court.

The Court's decision is not considered final or binding until it is announced by the Court. Until 1965, Decision Day—the day when decisions are announced—was only on Mondays, but now other days are also used for announcement of decisions. Justices once read extended portions of their opinions in open court. This practice has been eliminated in favor of short statements of what the case involves and what the Court has decided, in order, as Chief Justice Burger has said, to gain "some valuable time at the expense of pleasant, traditional, but unproductive ceremony."[1]

The Chief Justice plays a particularly important role in the Court's disposition of cases. This stems not only from his screening of certiorari petitions and his conference and opinion-assignment tasks, which provide only part of his authority. If he wants to lead the Court, as Chief Justice John Marshall was able to do in replacing the practice of each justice writing serial opinions with an "opinion of the Court" behind which he "massed" all the judges, he must use interpersonal influence as well. A Chief Justice devoting much attention to matters outside the Court such as judicial administration is less likely to be the in-court leader than if he invests most of his energies in the Court itself.

The Chief Justice may, like Taft, be the Court's social leader, the one who helps produce solidarity among the justices and "attends to the emotional needs of justices as individuals," or he may be its task, or intellectual, leader who "concentrates on the Court's decision."[2] Some Chief Justices, for example, Hughes, perform both roles, but few individuals have the skills necessary for both types of leadership. It is also possible that, as in the case of Chief Justice Stone, the Chief Justice will perform neither role, resulting in an extremely uncohesive Court. If the Chief Justice does not perform one of the necessary roles, someone else will; for example, Justice Willis Van Devanter was task leader when Taft was social leader. Even if the Chief Justice does not perform either role for the Court as a whole, he can perform either or both for a bloc of justices, as Chief Justice Burger apparently has done for the Nixon appointees.

An example of the Chief Justice's leadership is provided by Earl Warren's role in *Brown v. Board of Education*. The desegregation cases had already been argued before Warren joined the Court upon Chief Justice Vinson's death. At that time, a majority of the justices apparently already favored a ruling desegregating the schools, but there were two or three "holdouts" and the justices differed as well as to how to implement desegregation. Warren was largely responsible for the Court's unanimity in both its 1954 decision on the validity of desegregation and its 1955 "with all deliberate speed" ruling on implementation, and for the Court's opinions, which closely paralleled his original memoranda. In conference after oral argument he made clear that he strongly supported desegregation but, to avoid polarization among the justices, at first avoided taking a vote on the case. His "low-key approach, emphasizing fairness, understanding, and tolerance, combined with a strong plea for justice, clearly contributed to keeping the question on a mature level of discussion and to muting the differences (minor and major) among the justices."[3]

ORAL ARGUMENT

The Court sits from 10:00 A.M. to 2:30 P.M., with a half-hour for lunch, when hearing oral argument. Oral argument in a case once went on almost interminably. However, now there are time limits, an hour for each side until recently and now thirty minutes per side unless a case is exceptional, but consolidation of several cases for argument may mean more time on the basic issues common to them even if not on the specific facts of each case. Oral argument was used long before written briefs and is now used only in the most important cases. Although it may not determine the result, oral argument is often quite important in the resolution of those cases.[4] Justice Harlan stated that there was "no substitute" for this "Socratic method of procedure in getting at the heart of an issue and in finding out where the truth lies," and Justice Brennan said he would be "terribly concerned" were he to be denied the opportunity to participate in oral argument because there had been "many occasions when my judgment of a decision has turned on what happened" there.[5]

Particularly important at oral argument are the justices' questions. Lawyers do not make uninterrupted speeches but frequently engage in exchanges with the justices. Some justices are more frequent questioners than others, and the Court as a whole usually asks more questions of one side than of the other. For example, in the *Briggs* school desegregation case argued along with *Brown*, John Davis, defending segregation, was interrupted only eleven times, but Thurgood Marshall was interrupted 127 times.[6] The justices not only ask questions; they make statements and suggest positions not raised by the lawyers. In so doing, they often disagree with the lawyers, at times forcing them to back down. When the government's lawyer in the Pentagon Papers case complained that his client had been forced to prepare its case too hastily, Justice Stewart responded, "The reason is, of course, as you know, . . . that unless the constitutional law as it now exists is changed, a prior restraint of publication by a newspaper is presumptively unconstitutional."[7] And in the Little Rock case, when the state's lawyer tried to gain sympathy for Governor Faubus's position by recalling Chief Justice Warren's service as governor of California, Warren forcefully pointed out that he had as governor abided by rulings of the courts, adding, "I never heard a lawyer say that the statement of a governor as to what was legal or illegal should control the action of any court."[8]

Oral argument performs a variety of functions. Because there is no assurance that the justices have actually read the briefs although it is quite likely that they have done so, it serves to assure the lawyers—and through them their clients—that the Court has actually heard the case. It also helps both lawyers and judges by forcing the lawyers to focus on the arguments they consider most important. Judges' questions quickly lead to a separation of central from collateral issues, a distinction often not discernible from the lawyers' briefs, which are organized point after point in correspondence to the facts and statutes involved in the case, and which often intentionally lack emphasis as the lawyer, never knowing which argument(s) might be persuasive, tries to convince the Court in as many ways as possible.

For the judges, for whom it is probably more important, oral argument not only legitimizes their judicial function but can also be used to obtain support for their positions or to assure them about an outcome toward which they are already inclined. Judges also use oral argument to communicate with their colleagues, asking questions of counsel which are of greater concern to their brethren than to themselves and debating each other through those questions. A justice hammering on a particularly difficult point may be trying to persuade his colleagues that they will have to resolve that point to decide the case in a certain direction.

Oral argument is basically intended to provide the judges with information. Sometimes they simply want to clarify the lawyers' positions; for example, the differences in rates at which Marshall and Davis were asked questions in the *Briggs* argument can be explained in large part by that need for clarification. "Davis was usually so unmistakably clear that the justices did not need to ask him to restate an idea," while "Marshall was questioned because his occasional oral 'shorthand' (as Chief Justice Vinson once called it) puzzled someone on the bench." This difference was reinforced by "Marshall's attitude toward the Court's questions and his mode of answer [which] provided a vivid contrast, . . . invit[ing] questioning," while Davis's responses to questions "discouraged questioning."[9] However, the particular frequency with which Marshall's 1953 argument attacking the "separate but equal" doctrine was interrupted—fifty-three times in roughly three-quarters of an hour—also resulted from the fact that it was the central part of his case and from his position as the person wishing the Court to adopt a major new position.

Oral argument also can provide judges with information to assist them in determining the Court's strategies, that is, how the Court should exercise its broader political role. Questions as to how many people might be affected by a decision and where the Court might be heading if it decided a case a certain way help elicit this type of information. As Leon Friedman has noted, "The Justices' purpose in oral argument is to force the lawyers to think out the political, social, and constitutional implications of their arguments." In so doing, "the Justices constantly seek to relate the immediate problem to analogous situations . . . and try to cut through the rhetoric of the lawyers to see the concrete result of any argument advanced."[10]

CONSENSUS AND DISSENSUS IN THE COURT

"Judicial decision-making involves, at bottom, a choice between competing values by fallible, pragmatic, and at times nonrational men engaged in a highly complex process in a very human setting."[11]

Those who say the Court finds law say differences among judges result from intellectual processes of reasoning about past doctrine. Although relatively few scholars still hold such a position, others with an interest in the work of the Court still at least talk as if they expect judges to put aside their personal values and to engage in the deliberate, logical consideration of cases, with results affected only by the facts of a case and the relevant precedents. Thus when a justice's decisions seem fully controlled by his attitudes, sharp criticism is quite likely to arise. Shortly before he left the Court, Justice Douglas was the target of such criticism, not for his civil liberties decisions —as predictable in the liberal direction as Justice Rehnquist's are in the conservative direction—but for his tax rulings. The record showed that although Douglas supported the taxpayer less than did the Court as a whole prior to 1942, after then his support of the taxpayer was very much higher than the Court's average. In 1943–1959 the taxpayer won one-fourth of the time, but Douglas voted for the taxpayer in 47 percent of the cases. In the next five years Douglas's support increased to 73 percent, while the taxpayer was winning only 17 percent of the cases. Douglas supported the taxpayer in 59 percent of his 1964–1973 decisions, well over the taxpayer's 26 percent "win rate." [12] Such criticism indicates the difficulty people have in accepting the fact that differences among justices may be a product of interpersonal relations or the justices' deep-seated attitudes about their judicial role or, more particularly, about policy. Were this not the case, shifts in the Court's direction would not be—as they are in fact—directly related to changes in "clusters" of judges resulting from shifts in personnel, with the justices' interpersonal relations either softening or reinforcing these differences.

Unanimity and Dissent

Particularly disturbing to those who feel that the justices should find the law is the Court's lack of unanimity, a further indication that the Court makes policy. However, even those who accept the Court's policy-making role feel there is strength in unanimity and weakness in disagreement, most notably when the Court produces only plurality opinions—as it did in eleven cases (including the five death penalty cases) in the 1975 Term. This is likely to produce criticism even from within the Court, as when the Court's division over obscenity during the 1960s led Justice Clark to say, with some irritation, that his colleagues were "like ancient Gaul. . .split into three parts."[13] Such criticism does serve as a constraint on disagreement. A justice contemplating a dissent must also keep in mind that a dissenting opinion, particularly a strong one, may make the majority opinion more visible—and thus more damaging—than if he had remained silent. Furthermore, open disagreement may be interpreted as an attack on the Court's integrity. Such considerations of institutional loyalty do not, however, prevent dissenters from at times directly attacking not only the majority's policy positions but also its intelligence and ability.

Objection to internal division is based in part on the idea that dissents encourage noncompliance and that compliance is more likely to result from a unanimous than from a nonunanimous decision. However, minority opinions, in addition to being the potential majority opinions of the future, may give the losing side the feeling that someone has listened. They may also make the majority sharpen its reasoning and may allow the majority judges to write a more forceful opinion because they do not have to compromise with the dissenters as they would if the latter were still part of the majority.

Despite year-to-year variation, the level of disagreement within the Court has generally increased over time. Although "deep cases . . . draw blood," most of the Court's work in the late nineteenth century "was not made up of great cases," and as a result most cases—for example, all but five of seventy cases from the Court's 1866 Term—were decided unanimously.[14] As noted earlier, the Court's exercise of its certiorari jurisdiction led to a substantial increase in dissents. In 1930 only 11 percent of the Court's cases were split decision. However, that percentage more than doubled in each of the next two decades, to 28 percent in 1940 and 61 percent in 1950. After Earl Warren became Chief Justice, the figure rose to over three-fourths of the cases in 1957.

Only about two-fifths of the Court's full opinion cases in the early 1960s were unanimous. That figure dropped to one-third in the Warren Court's last term, but relatively few of the cases in the 1967 and 1968 terms were decided by close votes, with at least four votes separating majority and minority in roughly three-fourths of the cases. Moreover, individual judges' dissent rates were low, although Justice Harlan was a particularly frequent dissenter during the Warren Court's more liberal years. Toward the end of the Warren Court, Warren himself, Brennan, and Marshall seldom dissented, and in the last Warren Court term (1968), four justices had fewer than ten dissenting votes and only five justices cast any "solo" dissents.

The transition to the Burger Court produced an obvious drop in consensus. In Burger's first three terms, only roughly one-third of the full opinion cases were unanimous. This dropped to one-fourth in the 1972 Term and was only 29 percent in the 1974 Term before increasing to 36 percent in the 1975 Term.[15] The size of majorities reflected disagreement even more clearly. During the new Chief Justice's first term, there were wide margins in 62 percent of the cases; the next year, however, this was true in 48 percent, with one-vote margins in over one-fourth of the signed full opinion cases. The proportion of cases decided by wide margins increased to over half as the Court's membership stabilized after 1971, but one-fifth of the cases in the 1972 through 1974 Terms were decided by only one vote. In the 1975 Term, with the new Court more sure of itself, over 70 percent of the cases were decided by margins of at least four votes and only 10 percent by one-vote margins. However, the number of concurring opinions, which had increased in frequency from 40 in the 1962 Term of the Warren Court to 67 in the 1972 Burger Court Term,[16] did not abate, reaching a high of 85—with 115 concurring votes—in the 1975 Term.

In considerable contrast with his predecessor, in his first term Chief Justice Burger led the dissenters, departing from the majority 27 times, more frequently than either Black or Douglas. This was a clear indication of his lack of interest in massing justices behind a single opinion; for him, ideology was more important than solidarity. Even if we discount Justice Douglas's extremely high rate of dissent—57 dissenting votes in the 1971 Term (15 solo) and 69 the following Term—the overall rate of dissent in the Court was high, with over 200 dissenting votes cast in signed cases in Burger's second term. As the Nixon bloc came to control the Court, the liberals became the most frequent dissenters, and Burger's rate of dissent decreased, although Justice Rehnquist established himself as a frequent dissenter among the Nixon appointees.

Even though over time the new majority's increasingly sure control of the Court reduced the number of cases decided with close votes, the numbers of dissents remained high. Thus in the 1975 Term there were over 220 dissenting votes cast and over 110 dissenting opinions written. Part of the reason for this dissension is that the newer members of the Court knew their own minds and did not wait to assert their positions. The suggestions that justices "bide their time" before assuming a position at either the liberal or conservative side of the Court, perhaps staying "in the middle" for several terms before "making their move," and that each waits cautiously to cast his first dissent in a case of substantial importance have not applied to the recent justices. Perhaps as a result of their past judicial and legal experience, they have spoken out in dissenting and concurring opinions within a short time of reaching the bench. Thus Justice Stevens, despite his nonparticipation in roughly one-third of the cases in his first term, dissented in nineteen cases, writing opinions in sixteen—more than all the other justices but Brennan and Marshall.

Part of the difference in agreement and disagreement patterns between the Warren and Burger Courts may stem from the presence or absence of external threat. Those who study small-group decision making say that "minimum winning coalitions," that is, the smallest majority necessary, will make decisions in such groups, particularly when they are not threatened from outside. Supreme Court votes seem to support that proposition. The Burger Court was not so threatened, although when President Nixon indicated doubt and hesitation about complying with judicial orders relative to subpoenas for his tapes, the Supreme Court ruled against him unanimously (8-0). However, in the Warren Court, which was threatened by legislative proposals to limit its jurisdiction, in threat situations only 13 percent of the decision coalitions were minimum-winning (31 percent, combining five- and six-person coalitions), with 44 percent of the decision coalitions being unanimous, while in nonthreat situations, 31 percent were minimum-winning (56 percent, combining five- and six-member coalitions), with only 17 percent of the decision coalitions unanimous. For opinion coalitions, that is, those joining the majority opinion as well as the decision, the figures were somewhat different. In threat situations, 23 percent of the coalitions were minimum-winning, in nonthreat situations, 40 percent. (With five- and six-member coalitions combined, the figures are 45 percent and 63 percent.) However, most nonthreat minimum-winning opinion coalitions resulted from minimum-winning decision coalitions, and decision and opinion coalitions were identical in 69 percent of the nonthreat cases, suggesting that threat does not necessarily increase opinion coalitions. Furthermore, when coalitions were large under conditions of threat, they were more likely not to be "connected," that is, to include justices whose policy positions were close together. When the opinion coalition was smaller than the decision coalition, the coalition was more likely to be "connected."[17]

Judicial Alignments

Studies of the justices' values or attitudes were not likely to have been carried out by those who believed that judges only found law. Once the role of judges' values was admitted, however, such studies were possible. The first such major systematic study was Pritchett's examination of the Roosevelt Court based on justices' agreement with each other in nonunanimous cases.[18] It was followed by other studies, primarily of the Warren Court.[19] While most studies have been based only on the justices' recorded votes, materials from outside the Court have at times been used. For example, Danelski hypothesized that a judge's previous experience with respect to a value—such as morality, tradition, or religion—could affect his perceptions in terms of that value. He looked at the relationship between the values present in speeches men had made before becoming Supreme Court justices and their judicial positions as reflected in their solo dissents. In addition to finding a congruence between the values expressed in the two types of statements, Danelski also found that judges held values which caused internal conflict for them. For Justice Pierce Butler, these noncongruent values were patriotism and individual freedom. Butler resolved the conflict between them by deciding in favor of patriotism in free speech cases but in favor of individual freedom in criminal procedure cases.[20]

The voting studies of the Court have been based primarily on the justices' voting interagreement (bloc analysis) and on Guttmann scaling, a technique for testing the consistency of attitudes underlying a set of decisions, with some scholars using still more sophisticated methods.[21] Most bloc analysis studies have been based on the voting interagreement of pairs of justices. The pairs are arranged in a matrix, with blocs identified from sets of high interagreement scores. However, if a bloc is defined as a cohesive group whose members vote together regularly, so that a "bloc" really exists only if Justices Jones, Green, and Brown vote together frequently, and if the three are frequently found alone, bloc behavior may not be very frequent. If they vote together primarily when they are also voting with others, they do not constitute a bloc. Similarly, the members of several pairs of justices do not constitute a bloc unless all the members of the pairs in the set are in high interagreement. For the 1963–1965 Terms, for example, it was difficult to determine the identity of blocs which persisted over time, and identifiable blocs "voted together alone less than half the number of times [they] voted together." Moreover, three- and four-judge dissenting blocs seldom joined in the same *opinion* even when they voted together making tenuous the idea of "stable, persistent and exclusive. . . blocs, whose members interact substantially as a bloc."[22]

Whatever their methods, researchers have regularly discovered blocs in the Supreme Court. For example, for the 1931–1935 Terms, there was a "Left bloc" of Stone, Cardozo, and Brandeis, and a Van Devanter-Butler-Sutherland-McReynolds "Right bloc." Justice Roberts and Chief Justice Hughes were members of the Right bloc in terms of overall interagreement, although in dissent Roberts was independent and Hughes was marginally affiliated with the Left bloc. Extending this analyses, Schubert calculated what Hughes and Roberts ("Hughberts") would have to have done in the 1936 Term to increase their power, that is, their ability to determine the outcome of the Court's decisions.[23] A "pure" strategy for them would have been to have formed a minimum-winning coalition of five with the Left bloc when possible, to join with the Right bloc when the first strategy was not possible, and to join the rest of the Court (Right and Left together) when all agreed, so as not to be isolated. The Hughes/Roberts behavior indeed closely approximated the specified strategy: the two were affiliated with the Left in total interagreement, voting with the Right only when they could not form a majority with the Left.

Part of the "conventional wisdom" concerning the Court's response to FDR's attack was that only Roberts had shifted his position to produce the "switch in time that saved nine." Roberts's own explanation was that he had earlier indicated his change in positions on the validity of minimum wage legislation. However, when a case containing the appropriate challenge did arise, Justice Stone's illness produced a delay in decision of the case, which as a result was not decided until after the Court-packing plan.[24] Schubert's analysis, which led him to conclude that both Roberts *and* Hughes had shifted, contradicts both explanations but is reinforced by a radical shift in bloc dissenting rates from the five previous terms to the 1936 Term. Furthermore, the study covers an entire term of Court, not just a few cases, and reveals patterned rather than idiosyncratic behavior.

Another bloc which has been identified is a "Truman bloc" in the late 1940s consisting of Truman appointees Vinson, Clark, Burton, and Minton and Roosevelt appointee Reed. During the 1946–1948 Terms, in terms of overall interagreement, Reed, Burton, and Chief Justice Vinson formed a bloc in the center of the Court, while Justices Jackson and Frankfurter formed a bloc on the right of the Court. Burton joined Vinson and Reed in a Right bloc and also joined Frankfurter and Jackson in a center bloc when dissenting. There was a five-judge conservative majority on the Court except when the Reed-Vinson-Burton trio split, which made liberal majorities possible. In the 1949–1952 Terms, Black and Douglas were isolated on the left of the Court and Jackson and Frankfurter were between them and the five-member Truman bloc. It has been suggested that in the 1950–1952 Terms, the Truman bloc was largely influenced in its voting by a desire to protect the Court's institutional stability. However, the voting by Burton in the 1950 Term, Reed in the 1951 Term, and nonbloc member Jackson in the 1951 and 1952 Terms can be explained in terms of their wishing to avoid isolation. On the other hand, Frankfurter, Douglas, and Black voted in terms of their basic ideological positions for all three terms, as did Jackson in the 1950 Term.[25]

Justices' interagreement in the Warren Court and afterward indicates the presence of blocs as well as variance in the patterns of interaction among and within those groupings. In the 1961 Term, in 53 nonunanimous full opinion cases, Warren voted over 40 times each with Brennan, Black, and Douglas. The four voted together 28 times, and Warren was joined by two of the three on 11 other occasions in the majority and dissented with both Black and Douglas nine times. In the 1962 Term, Warren had high rates of interagreement with Brennan, Black, and Douglas (67, 56, and 58 of 71 nonunanimous cases, respectively) and the four voted together a full 48 times, all in the majority. Furthermore, the four together voted 40 times with Goldberg, the fifth liberal justice. Goldberg in turn paired with centrist Justice Stewart 46 times, eight of those times in dissent. Harlan and Stewart voted together 48 times, more than half in dissent, and Harlan also joined Clark and White regularly, also more in dissent than in the majority.

The 1963 Term again found Warren, Brennan, Black, Douglas, and Goldberg voting together regularly, with the highest levels of interagreement between Warren, Brennan, and Goldberg. Goldberg and Stewart continued to vote together more than half the time, and also persisting were the Stewart-Harlan-Clark-White alignments, although interagreement among the conservative justices was lower than among the liberals. In the 1964 Term, when the percentage of the Court's liberal decisions declined, Warren's interagreement with Brennan remained high, but his interagreement with Black, Douglas, and Goldberg decreased from previous terms. (Warren and Brennan agreed at levels over 90 percent throughout the remainder of the Warren Court.) In the 1965 Term, the Chief Justice's pairing with Black and Douglas was exceeded by his interagreement with new member Fortas and with Justices Clark and White; the 1966 Term showed little change from 1965.

For the mid–Warren Court period (1962–1964 Terms), Schubert has shown that interagreement patterns indicated an alignment of the justices from Douglas, the most liberal, through Black, Warren, Brennan, Goldberg, Clark, White, and Stewart to Harlan, the most conservative. The first five formed a liberal bloc, the remaining four a conservative one.[26] In the 1965 and 1966 Terms, when Fortas replaced Goldberg in the liberal bloc, Black had left that grouping, moving to a position midway between the two blocs.

While Black's increasingly conservative voting behavior may have resulted from aging and senscence, he may also have been responding to the new types of civil liberties cases facing the Court, involving not "pure" free speech but "free speech plus," particularly civil rights demonstrations—to which he had a strong negative reaction.[27]

In the 1967 and 1968 Terms, Douglas, continuing his liberal voting, moved to isolation left of the liberal bloc, which now included Thurgood Marshall. Black remained in isolation from both the liberal bloc and the White-Stewart-Harlan conservative bloc. Further indication of Black's movement away from the liberals is that he voted with Stewart as frequently as with Warren. The four-member grouping of Warren, Black, Douglas, and Brennan, often joined by Fortas and Marshall, voted together in only a third of the cases, but three of the four were together in two-thirds of the cases. Harlan joined White more frequently in the majority and Stewart more frequently when he dissented, but all three continued to vote together frequently, as they also did during the transitional 1969 Term when the Douglas-Brennan-Marshall trio, which extended into subsequent terms, began to be evident.

In 1969 Black voted with the new Chief Justice more than he did with former voting companion Douglas. Brennan and Marshall formed the strongest liberal pair, but the better-known pair was the Chief Justice and Justice Blackmun, the "Minnesota Twins." They voted together in 90 percent of the 1970 Term's nonunanimous cases and in all but one criminal procedure case. However, their agreement decreased somewhat, to only 80 percent, in the next term, when Burger's highest interagreement was with Rehnquist (95 percent). The 1970 Term also revealed frequent (70 percent) interagreement between White and Stewart, who were never in dissent together and who helped the Nixon appointees control the Court: the term's most frequent majority voting combinations were the Nixon Four plus Stewart and White or White alone. The sharp division between the new majority coalition and the remaining liberal justices was clear from Burger's agreement with Douglas in only 4.2 percent of the cases, Brennan in only 10.9 percent, and Marshall in only 21 percent.

The Nixon appointees' interagreement, while not perfect, was quite high. In the 1972 Term, typical of what was to follow, three of the four Nixon appointees were together in 85 percent of the nonunanimous cases. Burger, Blackmun, Powell, and Rehnquist were together in almost three-fifths of those cases, only once in dissent. With White and Stewart, they formed a majority in 25 cases; in 14 more cases, they did so with White alone. Goldman shows two blocs for the 1970 through 1972 and 1974 Terms, three for the 1973 Term. In the 1971 Term, Justice Stewart was to be found with the liberal bloc of Douglas, Brennan, and Marshall, but in 1973, Stewart, along with White and Nixon appointee Powell, formed a "midbloc" on the Court.[28]

The 1975 Term showed greater stability in the Court and greater assuredness from the conservatives. The most common pattern in nonunanimous cases remained the four Nixon appointees plus White and Stewart; the next most common involved those six justices plus Stevens, the Court's newest member. With Douglas gone, the most frequent dissenting pattern was the liberal pair of Brennan and Marshall, who voted alone as a pair 22 times; the two were joined five times each by White and Stevens. With almost 90 nonunanimous cases, the "Nixon Four" was together in 50 cases, never in dissent (although they were in the minority on one order to allow certain persons to proceed *in forma pauperis* in the Court). Three of the four Nixon appointees were together 79 times, only seven in dissent. High rates of agreement continued to be registered by the "Minnesota Twins" and by Burger and Rehnquist. White and Stewart also voted together frequently, but as might be expected of centrist judges, less frequently than the just-noted conservative pairs or the liberal Brennan-Marshall pairing. Showing his mid-Court position, new Justice Stevens paired at roughly the same rates with Burger, Stewart, White *and* Brennan.

The "defection" of one of the Nixon appointees did not necessarily deprive them of control of the Court. In 1972 Stewart and White voted with three of those four to form a majority in four cases. However, Blackmun and Powell each once became the deciding fifth vote in a liberal criminal procedure majority and the four also divided evenly in sixteen cases, with Blackmun and Powell usually on the more liberal side. These divisions occurred on such important issues as Denver school desegregation, sex-related newspaper employment advertisements, and tuition grants for parochial education. In the 1973 Term, when the four split 3-to-1, the three were in dissent in only eight of 25 cases, least likely in the criminal procedure area; in 1974, there were thirty 3-to-1 splits. In that term, Justice Blackmun, who became the member of "the four" most likely to leave them, was the Nixon appointee most likely to be in the majority in civil liberties cases, voting against the civil liberties claim 42 or 70 times (60 percent); Rehnquist and Burger were closer to 70 percent against the claims, with Powell (and White) closer to Blackmun's record. Although the Court had become more conservative, the judges most likely to be in the majority were those with only moderately rather than extremely negative civil liberties voting records.[29]

The 1975 Term did show seven cases in which 3-to-1 splits among the Nixon Four produced a loss of their control of the Court, with Powell (four cases) and Blackmun (three cases) the "defectors" in those instances, which included cases on government employment of aliens, patronage, standing in Medicaid abortion cases, seniority as a remedy for employment discrimination, and the death penalty. (In the six cases when the four split 2-to-2, Blackmun and Powell were together in four of the cases.) Despite such occasional division, over several terms when the Court was opposed to defendants' claims in criminal procedure cases, the Nixon bloc was in the majority in all but two cases—97 percent—but when the Court majority supported a defendant's claims, the bloc was in the majority only slightly more than half the time.[30]

Studies based on Guttman scaling have produced findings about the Court's alignments comparable to those derived from bloc analysis. Scales are sets of cases in which the justices were aligned so that a consistent underlying attitude accounted for their votes. The basic scales which regularly appeared were the "C" and "E" scales.[31] The C scale covers civil liberties cases, matters of personal rights and freedoms such as the First Amendment, fair trial, and racial equality, while the E scale includes economic regulation matters: government regulation of business, antitrust, and labor-management cases. In some terms of the Court, the C scale, rather than being a general civil liberties scale, has had subcomponents. In some years, other scales, including one encompassing questions of federalism and another judicial activism in reviewing decisions of the other branches of government, have also been identified.

By combining justices' positions on the political (C) and economic (E) scales, Schubert identified four categories of justices. The four types of judges were the Individualist, conservative in economics but politically liberal; the Liberal, liberal on both scales; the Conservative, conservative on both; and the Collectivist—an economic liberal but a political conservative. Further classification produced three ideological dimensions based on political and economic conservatism/liberalism, plus a fourth dimension—Pragmatism/Dogmatism—which cut across the others and reflected judges' beliefs in the authority of precedent (whether rules should be followed until formally changed). Dogmatists were those who believed highly in precedent; those less concerned about precedent and more concerned about decisions' effects were Pragmatists. In 1946–1962 conservatives were divided into Pragmatists and Dogmatists, but liberals were not.

One of the three basic dimensions was that of Equalitarianism/Traditionalism; someone liberal on both C and E scales would hold the Equalitarian ideology, a belief in greater equality of opportunity for all. The Authoritarian/Libertarian dimension dealt with the scope of freedom as opposed to the scope of authority; those holding the Authoritarian ideology were political conservatives and economic liberals. The third policy dimension was Collectivism/Individualism, involving emphasis on the individual human being as against emphasis on society as such, with a political liberal and economic conservative an Individualist. Using these categories, one could identify the dominant ideology in the court. Thus, Traditionalism was dominant in 1946–1948, Collectivism in 1949 and 1954 (both transitional periods), Authoritarianism in 1950–1953, and Equalitarianism in 1955–1962.

During this extended period, on the dimension of political liberalism seven justices—Frank Murphy and Wiley Rutledge from the early Vinson Court, Black, Douglas, Brennan. Goldberg, and Warren—were positive, six were neutral, and five, including Chief Justice Vinson and Justice Clark, were negative. On the E scale, six justices, the same as on the C scale except for Goldberg, were positive, eight neutral, and four negative: Jackson, Frankfurter, Harlan, and Whittaker. Thus six justices who "sponsored change in the Court's policy-making in the direction of support for both civil liberties and greater economic egalitarianism" were positive on both scales, and four, the "economic conservatives," were neutral as to political liberalism but negative as to economic liberalism. Their "support of civil liberties was sufficiently unzealous that, when the Court itself came under political attack in 1956 and 1957 because of its emerging zeal for political liberalism, they readily supported the third cluster justices in checking the liberals." That third cluster was composed of the five (the "political conservatives") who were neutral as to economic matters but negative on political liberalism.[32] Only three justices—Warren, who became more liberal after his first few terms, Black, and Frankfurter—changed ideological positions once they joined the Court. Because so few justices changed positions, shifts in domination of the Court had to come through changes in personnel which in turn altered the three clusters of judges.

Schubert shows that Frankfurter began his judicial career as an economic conservative but then became conservative on both economic and political dimensions. Others have preferred to take Frankfurter's statements of judicial self-restraint at face value as the appropriate explanation for his actions. A study by Grossman provided a test of these different possibilities. A number of cases in which Frankfurter participated, particularly on civil liberties and labor matters, also contained questions closely related to "self-restraint." These questions concerning whether the Supreme Court should decide a case on the merits included jurisdictional matters and issues such as deference to other units of government. Where such a Denial of Judicial Responsibility (DJR) factor was present, Frankfurter's vote was always against the civil liberties claim in civil liberties cases and against the liberal position in sixteen of eighteen labor cases. With the DJR-related cases removed from the civil liberties scale, Frankfurter's position relative to other judges remained to the right of center. Thus, even if the DJR factor "controlled" the civil liberties ideology for Frankfurter where the former appeared, in the other cases where it could not do so Frankfurter was obviously a conservative rather than an apostle of self-restraint.[33]

Another study, covering the Warren Court but also extending into the Burger Court, produced the finding that more than 85 percent of the Court's rulings could be explained in terms of three major dimensions, two related to civil liberties and one, New Dealism, based on economic activity. The first civil liberties dimension, freedom, covered cases involving criminal defendants and those in political "crimes," e.g., the loyalty oath cases; the other, equality, covered political, economic, or racial discrimination. Two other relatively minor dimensions—privacy (libel and obscenity) and taxation—also helped explain Warren Court rulings.[34] Six justices—Warren, Douglas, Brennan, Goldberg, Fortas, and Marshall—were liberal, that is, positive on the three major dimensions. Seven, the four Nixon appointees, Frankfurter, Whittaker, and Harlan, were conservatives, negative on all three. Justices Stewart and White were considered Moderates, Black was a Populist—negative as to equality, but positive as to freedom and New Dealism—and Clark was a New Dealer, positive only on economic matters, negative on the other two dimensions.

Guttmann scaling studies have also suggested whether the Court's decisions have been based on the justices' attitudes toward a situation (AS) or toward an object (AO), that is, whether they would vote in terms of their attitude toward that type of person regardless of the situation. In the Warren Court's last eleven terms, AS was found to be more determinative in cases involving labor unions and persons exercising freedom of communication, while AO was more important in security risk cases and physically injured employees.[35] For the 1971–1973 Terms, the attitude toward object —support for the individual or for the government—dominated situational variables for eight of the nine justices in both criminal and noncriminal cases. Only Justice Powell differed between criminal and noncriminal cases, supporting the government to a statistically significant degree in the former but not the latter. That the other Nixon appointees favored the government over the individual in both criminal and noncriminal situations indicated clearly that their voting resulted from broader ideology, not a narrower and isolated "law and order" stance.[36]

All these studies are helpful in understanding the Court and its members, but they are also limited. They show only justices' positions relative to each other, not in absolute terms. Thus we cannot tell from scale positions either where the judges on the extremes are going to fall or exactly where the division between majority and minority will occur. Because they are based only on nonunanimous cases, the scales also cannot explain why the Court achieves unanimous decisions not only on noncontroversial legal questions but on important matters of law such as school desegregation and the Nixon tapes. They show overwhelmingly that attitude and ideology affect justices' votes but not the other factors which also influence the result and which help to explain why not all justices can be located easily on all scales.

Perhaps most significant is that these studies are based only on votes rather than the justices' opinions and generally do not take into account the fact that justices who vote together do not necessarily write together. Because they capture only the Court's final product, its decision, such important matters as interaction among the justices and the changes in written opinion and even result which may occur between the initial conference vote and final announcement of a decision are lost. The result in a case as determined by the vote is perhaps the most important to the litigants, but the doctrine of an opinion is generally thought to be more important for the Court's role as a policy maker. Pragmatism, evangelism, and milieu of advocacy are among the factors which diminish the effect of attitude and ideology, particularly with respect to the opinions the justices write. Pragmatism involves the question of "strategic judgments about what professional and lay traffics would bear." It can be seen when a justice withholds a dissent even when opposed to the result or limits himself to a concurring opinion rather than a dissent, perhaps so that the Court's decision can have greater force through a larger majority. Evangelism, the attempt to persuade publics outside the Court of the correctness of the Court's doctrine, certainly influences not only the justices' votes but also what they say. Thus "the demands of persuading colleagues and countrymen in trailblazing cases coalesce with professional habits and personal antagonisms to transform opinion-writing into argument and over-statement."[37]

OPINION ASSIGNMENT

The Court's policy is most explicitly developed through its majority opinions, and the Court's basic impact on other government units comes through those opinions, which "determine the kinds of claims the justices themselves will subject to their attitudes in the future."[38] The breadth of an opinion, its doctrinal bases, and the other content it contains—including precedent and history, to establish continuity with the past, and social science materials, generally used sparingly—affect not only an opinion's immediate policy effects but also its value as precedent. Whether an opinion is framed narrowly, staying within the grounds urged by the parties or seemingly applicable only to the facts of the particular case, or departs from those facts to embrace other situations, and whether it is confined to a statute rather than reaching a constitutional question, are quite important. The Court may wish to avoid a broad opinion which looks more "legislative" as well as a ruling so narrow it does not illuminate the law for other lawyers and judges and thus leads to the filing of many more cases. However, narrow rulings may be useful as a delaying tactic or as a device to clear away the underbrush before an unmistakable direct attack is mounted on a particular practice. They may also be necessary because a broad opinion may also lose judges from the majority, with the ultimate opinion often the lowest common denominator the justices can produce.

For all these reasons, the assignment of opinions by the Chief Justice is one of the ways in which he can attempt to exercise his and his colleagues' influence outside the Court—on the Court's multiple audiences, particularly lawyers. Such considerations thus place constraints on his choice of the Court's opinion writer. So do his efforts to influence his colleagues through the assignment of opinions and the necessity of distributing the writing workload evenly among all the justices, something not easy to do if some colleagues are frequently in dissent. Justices are expected to be "generalists," that is, to write opinions in all fields of law, but informal specialization appears from time to time—for example, with Justice Stephen Field and land law and apparently with Justice Powell in business cases.[39] The Chief Justice must participate in writing opinions, and he thus retains some cases. It is expected that the Chief Justice will write in some of the Court's "big" cases, but the Court's internal social relations prevent him from keeping all the most important ones.

Danelski has suggested two strategies which the Chief Justice might use in assigning opinions to accommodate some of these pressures:

RULE 1: Assign the case to the Justice whose views are the closest to the dissenters on the ground that his opinion would take a middle approach upon which both majority and minority could agree.

RULE 2: Where there are blocs on the Court and a bloc splits, assign the case to a majority member of the dissenters' bloc on the ground that he would take a middle approach upon which both majority and minority could agree and that the minority Justices would be more likely to agree with him because of general mutuality of agreement.[40]

If we look at Chief Justice Warren's opinion assignments, we find that in close political cases through 1962, he "overassigned" cases to two categories of judges: the majority judges closest in position to the minority and the justices who joined the majority through votes inconsistent with their usual positions. He thus seemed to follow the strategies suggested by Danelski. In economic cases, however, opinions were assigned to maximize the liberal policy position.[41] Where majorities were large, Warren tended to overassign to those in the midmajority position. As had Chief Justice Vinson before him, he also underassigned to himself and used those large majorities to help create an opinion of the Court considerably different from his own particular position. This was the case although assigning justices have generally favored justices "closest to them in various issue areas."[42] Those justices never "closest" to the assigners are usually given unanimous cases or cases in less controversial areas of the law.

Through 1960 Warren also underassigned to Frankfurter overall but gave him and Black a larger proportion of significant cases and Douglas and Clark a smaller proportion. Frankfurter, Douglas, and Clark were also particularly likely to receive assignments in cases decided by 5–4 votes, of which the Chief Justice took very few for himself. Douglas was more likely, Frankfurter less likely, to write an economic policy.[43] The rule of assigning cases to moderate justices was supported by Clark's opinions in cases with liberal results, such as *Mapp v. Ohio* on the exclusionary rule for improperly seized evidence, where his background as attorney general was probably thought to produce a better reception in the law enforcement community than if a justice with a predictably liberal record had written for the Court.

In the Court's 1962 Term, Justice Goldberg, the fifth reliable liberal vote but the one closest to the Court's center, wrote six of his twelve opinions in 5–4 rulings. Justices Black, Stewart, and Clark wrote mostly in unanimous cases, as did the Chief Justice himself, although he also assigned himself two 5–4 rulings. The next term, in which Warren wrote the reapportionment opinions, showed the same pattern, with Goldberg writing four 6–3 decisions and the Court's 5–4 *Escobedo* opinion. Frequent dissenter Harlan wrote in unanimous or nearly unanimous cases, and White, also on the Court's conservative side, wrote in cases with wide margins. When Fortas replaced Goldberg in the 1965 Term, he wrote primarily in unanimous rather than close cases. In the 1967 and 1968 Terms at the Warren Court's end, Fortas and Brennan (in 1967) and Warren himself (in 1968) were more likely to write in unanimous cases. Warren also seemed to keep more of the Court's important cases than he had earlier, although he by no means monopolized them.

Chief Justice Burger was reported to have tried to control opinion assignment even when he was not in the majority, perhaps as a result of his dissenting more frequently than Chief Justice Warren had done. Apparently he was blocked from doing so, but instead of trying to control the final opinion itself, he may only have been trying to obtain an issue-clarifying memorandum in advance of the Court's opinion like those used in some other courts. While Warren had allocated workload evenly, Burger did not do so in his first few terms, although vacancies certainly hindered him from this. The number of opinions assigned to each judge smoothed out after Powell and Rehnquist joined the Court. However, in the 1971 Term and after, there were discrepancies in the rates at which the justices were assigned a case when they were available to the Chief Justice. In 1974, for example, Blackmun was used 12 percent and Marshall 12.3 percent of the time each was available to Burger, while Powell and Rehnquist were each assigned opinions in 16 percent of the cases when they were available; the

Chief Justice self-assigned 13.3 percent of the time. Such percentage differences, while necessary to help equalize the number of opinions, did not always produce that result. For example, in the 1974 Term, seven justices each wrote from thirteen to seventeen opinions of the Court, but Justice Marshall wrote only ten, while his fellow liberal Brennan, who dissented slightly more, wrote fourteen. In the following term (1975), however, Marshall wrote more opinions than any other member of the Court, being assigned to do so in 21 percent of the cases in which he was available to the Chief Justice.

Burger underassigned to both Brennan and Harlan in his first term but used Douglas regularly despite the latter's relative unavailability because of his frequent dissents. Of note was that Justice White wrote for the Court in eight split-opinion cases, particularly in the criminal procedure area. The following year (1970), the Chief Justice relied particularly on Black and Stewart and underassigned to Justice Blackmun, whose first term it was. Doublas and Brennan wrote primarily in unanimous cases that year, while Black—given the crucial *Younger v. Harris* injunction cases—and White wrote principally in close cases, reinforced by Black's assigning himself three 5–4 decisions. (Douglas, in assigning opinions, overassigned to Justice Brennan and himself, particularly in the 1971 and 1972 Terms.)

Through his assignment of opinions, Burger definitely rewarded the other Nixon appointees, and to a somewhat lesser extent, White and Stewart. White was used heavily in the criminal procedure area; so was Rehnquist, whose assignment ran counter to the strategy of using mid-Court members, but may have resulted either from the major coalition's single-mindedness in this policy area or from a feeling that a more moderate justice would not attract more votes or that the coalition was sufficiently solid that Rehnquist's position would not lose votes. In the 1974 Term, both Rehnquist and Powell wrote heavily in 6–3 and 5–4 cases, Powell writing in five 5–4 and three 6–3 rulings, In 1975 Rehnquist wrote five 5–3 opinions, while Blackmun and Powell wrote predominantly in 7–2, 6–2, and 6–3 decisions. New Court member Stevens wrote only eight opinions, but half were 5–4 decisions, two assigned by Brennan. Except for Brennan's opinion in the Denver school desegregation case, in 1974 he, Marshall, and Douglas received less important cases or wrote when the Court was unanimous, a pattern continued for Brennan and Marshall in 1975 after Douglas' departure.

Burger assigned himself more major cases than had Warren, although in Burger's first two terms he took most unanimous ones, as he was to do again in 1975. Where Warren had written often in cases where the vote was unanimous or the vote margin was wide. Burger did not hesitate to write in close cases, for example, in nine 5–4 rulings in the 1972 Term. Burger's opinions over the first five terms included the *Swann* and Detroit busing cases, plea bargaining, school attendance by the Amish, the major obscenity case of *Miller v. California*, access to the media and the Nixon tapes case.

OPINION RELEASE

Once a case is decided and the opinions written, they must be released. The Court's invariable practice has been to release them to the news media without additional comment. The Court has an Information Officer, but his basic task is to see that the news media get the opinions, not to explain them nor to respond to questions about what the Court or individual justices may have intended. Under Chief Justice Warren, the Court did move its public sessions from the noon-to-4:30 P.M. period to their present 10:00 A.M.-to-2:30 P.M. time slot to help the media meet deadlines; that, however, only provided a few more minutes for reporters to digest the rulings unassisted. The Burger Court has assisted the press further by having the case "syllabus" or headnote, which summarizes the facts and issues and sets out the Court's holding, accompany every case when the case is issued instead of having it prepared only months later for the official reports. While written by the Reporter of Decisions and not officially part of the opinion, the headnote is a concise summary and definitely assists the media.

The Court's former practice of using only Monday for Decision Day meant that large numbers of important full opinion cases—not to speak of certiorari denials and other orders—were handed down on the same day particularly at the end of the term, thus deluging the media. When the Court first shifted to using non-Mondays as Decision Days—the 1965 Term was the first full one with the new practice—one-fourth (25.8 percent) of the Court's full opinion cases were issued on days other than Mondays.[44] However, in the remaining three Warren Court terms, the Court made only minimal use of non-Mondays: 6 percent (1966 Term), 11.8 percent (1967), and 15 percent (1968). Moreover, toward the end of each term when spreading cases out might have helped most, Monday remained the only Decision Day. Thus there were such unusually high outputs as twelve (June 10, 1968), thirteen (June 12, 1967), and fourteen (May 20, 1968), many of which announced significant constitutional doctrine.

Chief Justice Burger definitely changed output patterns in this respect. In his first term, non-Mondays were used for 24 percent of the Court's full opinion decisions. This rose to 39 percent for the 1970 Term and then again to just under half before rising to two-thirds of all signed full opinion cases in the 1975 Term. Moreover, the Court developed a twice-a-week pattern for the end of the term, in 1975 even overcoming Monday's predominance at that time in the term. In any event, the Court succeeded in spreading out the cases more when the flow was heaviest. For example, a total of twenty signed opinions and one substantive *per curiam* ruling were handed down during the week of June 23, 1975, but no more than six opinions were issued on any one day during that week. In fact, except for the end of the 1972 Term, with one day of twelve opinions and two each of fourteen opinions, generally no single day produced a glut of cases; eight or nine seemed to be the new upper limit for a single day's output.

The Chief Justice did not, however, smooth out the flow of cases through the term. Few opinions can be expected in October, November, and December when oral argument has just begun, but disparities in output between the second three months (January–March) and the last three (April–June) have been considerable. In Burger's second term (1970), by the end of December only six full opinion rulings had appeared, and in the 1974 Term only 17 of 145 signed opinions and substantive *per curiam* rulings "came down" in the first three months. This was worse than the Warren Court's last four terms in which somewhat less than one-fifth of the Court's full opinions had been handed down in the October–December period. Of the last four Warren Court terms, only in the 1965 Term were over half the full opinions released by the end of March; in the 1967 Term, only 39 percent of the term's full opinion output had been announced by that date. Except for the 1974 Term when half the opinions were announced by March 31, this pattern continued with the Burger Court. In both the Warren Court and the Burger Court—except for the low output year of 1969, caused by a vacancy on the Court—as much as one-third of the Court's entire output for the term would be announced in the last six weeks, with the figure as high as 43.6 percent (1967 term) and rising to 40 percent or more regularly.

We have now seen, in the two previous chapters and this one, the rules that control the access of cases to the courts and the relations between federal and state courts, the considerations that affect the bringing of appeals within the judicial system, the Supreme Court's "winnowing" of cases brought to it, and, finally, the procedures it uses to dispose of the cases it considers most important, the ones to which it gives full-dress treatment. The divisions among the justices that develop in such cases have also been examined, as has the assignment of opinions by the Chief Justice and the Court's release of opinions. That release of opinions begins the process by which the Court affects the political world around it. In Part Four, we turn to examine that impact—first in terms of the Court's exercise of judicial review at the national level, with respect to Congress and the executive, and then in the local community, where we give particular attention to the communication of its decisions, to the impact of its criminal procedure and church-state rulings, and to some of the factors that affect the impact of the decisions.

NOTES

1. Chief Justice Burger to Senator Hruska, May 29, 1975, in Commission on Revision of the Federal Court Appellate System, *Structure and Internal Procedures* (Washington, D.C., 1975), p. A-224.

2. David Danelski, "The Influence of the Chief Justice in the Decisional Process," *Courts, Judges, and Politics,* edited by Walter F. Murphy and C. Herman Pritchett (New York: Random House, 1961), p. 498.

3. S. Sidney Ulmer, "Earl Warren and the Brown Decision," *Journal of Politics* 33 (August 1971): 700. See also Daniel Berman, *Is It So Ordered* (New York: W. W. Norton, 1966).

4. This section draws on Stephen L. Wasby, Anthony A. D'Amato, and Rosemary Metrailer, "The Functions of Oral Argument in the U.S. Supreme Court," *Quarterly Journal of Speech* 62 (December 1976): 410–22.

5. Quoted in Commission on Revision of the Federal Court Appellate System, *Structure and Internal Procedures,* pp. 104–105.

6. Milton Dickens and Ruth E. Schwartz, "Oral Argument Before the Supreme Court: Marshall v. Davis in the School Segregation Cases," *Quarterly Journal of Speech* 57 (February 1971): 39.

7. James F. Simon, *In His Own Image: The Supreme Court in Richard Nixon's America* (New York: David McKay, 1973), pp. 208–209.

8. Anthony Lewis, "Supreme Court of the Land Still Rests in High Court," *Portland Oregonian,* July 14, 1974, p. F3. For other examples from the race relations area, see *Argument: The Oral Argument Before the Supreme Court in Brown v. Board of Education of Topeka, 1952–1955,* edited by Leon Friedman (New York: Chelsea House, 1969). Oral argument in a series of obscenity cases is available in *Obscenity: The Complete Oral Arguments Before the Supreme Court in the Major Obscenity Cases,* edited by Leon Friedman (New York: Chelsea House, 1970).

9. Dickens and Schwartz, "Oral Argument Before the Supreme Court," pp. 36–37.

10. Leon Friedman, "Introduction," *Argument,* p. vii.

11. A law clerk to a Supreme Court justice, quoted by Nina Totenberg, "Behind the Marble, Beneath the Robes," *New York Times Magazine,* March 16, 1975, p. 15.

12. Bernard Wolfman, Jonathan L. F. Silver, and Marjorie A. Silver, *Dissent Without Opinion: The Behavior of Justice William O. Douglas in Tax Cases* (Philadelphia: University of Pennsylvania Press, 1975).

13. *Manual Enterprises v. Day,* 370 U.S. 478 at 519 (1962).

14. Lawrence Friedman, *A History of American Law* (New York: Simon & Schuster, 1973), p. 331.

15. Data from *Harvard Law Review* show even lower percentages of unanimous decisions, because they consider a concurring vote to break unanimity. In the 1969 Term, their figure was 28.7 percent unanimous, and it fell all the way to 18.9 percent in 1970, rising slightly to 22.5 percent in 1971 and 21.3 percent in 1972. Figures for the 1973 and 1974 terms were roughly the same—24.8 percent (1973 Term), somewhat higher, and 21.0 percent (1974).

16. Erwin Griswold, "The Supreme Court's Case Load: Civil Rights and Other Problems," *University of Illinois Law Forum* (1973): 624.

17. David W. Rohde, "Policy Goals and Opinion Coalitions in the Supreme Court," *Midwest Journal of Political Science* 16 (May 1972): 218–19; Michael W. Giles, "Equivalent Versus Minimum Winning Opinion Coalition Size: A Test of Two Hypotheses," *American Journal of Political Science* 21 (May 1977), 405–8.

18. C. Herman Pritchett, *The Roosevelt Court: A Study in Judicial Politics and Values, 1937–1947* (New York: Macmillan, 1948).

19. An example of a study of an earlier period is John D. Sprague, *Voting Patterns of the United States Supreme Court* (Indianapolis: Bobbs-Merrill, 1968).

20. David Danelski, "Values as Variables in Judicial Decision-Making: Notes Toward a Theory," *Vanderbilt Law Review* 19 (1966): 721–40; Danelski, *A Supreme Court Justice Is Appointed* (New York: Random House, 1964).

21. These include factor analysis and "smallest space analysis." See Glendon Schubert, *The Judicial Mind* (Evanston, Ill.: Northwestern University Press, 1965). For discriminant function analysis, see S. Sidney Ulmer, "The Discriminant Function and a Theoretical Context for Its Use in Estimating the Votes of Judges," *Frontiers of Judicial Research*, edited by Joel B. Grossman and Joseph Tanenhaus (New York: John Wiley, 1969), pp. 335–69.

22. Joel B. Grossman, "Dissenting Blocs on the Warren Court: A Study in Judicial Role Behavior," *Journal of Politics* 30 (November 1968): 1083, 1089.

23. Glendon Schubert, *Quantitative Analysis of Judicial Behavior* (Glencoe, Ill.: Free Press, 1959), pp. 192–210.

24. For Roberts's explanation, see Glendon Schubert, *Constitutional Politics* (New York: Holt, Rinehart, and Winston, 1960), pp. 168–71.

25. David N. Atkinson and Dale A. Neuman, "Toward a Cost Theory of Judicial Alignments: The Case of the Truman Bloc," *Midwest Journal of Political Science* 13 (May 1969): 271–83.

26. Glendon Schubert, *The Constitutional Polity* (Boston: Boston University Press, 1970), pp. 124–25. See also S. Sidney Ulmer, "Toward a Theory of Sub-Group Formation in the United States Supreme Court," *Journal of Politics* 27 (1965): 133–52.

27. S. Sidney Ulmer, "The Longitudinal Behavior of Hugo Lafayette Black: Parabolic Support for Civil Liberties, 1937–1971," *Florida State University Law Review* 1 (Winter 1974) 131–53. For other studies of individual justices, see Ira H. Carmen, "One Civil Libertarian Among Many: The Case of Mr. Justice Goldberg," *Michigan Law Review* 65 (December 1966): 301–36, and H. Frank Way, "The Study of Judicial Attitudes: The Case of Mr. Justice Douglas," *Western Political Quarterly* 24 (March 1971): 12–23.

28. Sheldon Goldman and Thomas Jahnige, *The Federal Courts as a Political System*, 2nd ed. (New York: Harper and Row, 1976), pp. 162–63.

29. Paul Bender, "The Reluctant Court," *Civil Liberties Review* 2 (Fall 1975): 87–88. For the 1976 Term, see Stephen L. Wasby, "Certain Conservatism or Surprise: Civil Liberties in the 1976 Term," *Civil Liberties Review*, 4 (October/November 1977).

30. David W. Rohde and Harold J. Spaeth, *Supreme Court Decision-Making* (San Francisco: W. H. Freeman, 1975), p. 109.

31. What follows is drawn primarily from Schubert, *The Judicial Mind*.

32. Glendon Schubert, *The Judicial Mind Revisited: Psychometric Analysis of Supreme Court Ideology* (New York: Oxford University Press, 1974), summarized in "The Judicial Mind Reappraised," *Jurimetrics Journal* 15 (Summer 1975): 279.

33. Joel B. Grossman, "Role-Playing and the Analysis of Judicial Behavior: The Case of Mr. Justice Frankfurter," *Journal of Public Law* 11 (1962): 285–309. See also Harold J. Spaeth, "The Judicial Restraint of Mr. Justice Frankfurter— Myth or Reality," *Midwest Journal of Political Science* 8 (February 1964): 22–38.

34. Rohde and Spaeth, *Supreme Court Decision-Making*, pp. 137–38.

35. Harold J. Spaeth et al., "Is Justice Blind?: An Empirical Investigation of a Normative Ideal," *Law & Society Review* 7 (Fall 1972): 119–37.

36. S. Sidney Ulmer and John A. Stookey, "Nixon's Legacy to the Supreme Court: A Statistical Analysis of Judicial Behavior," *Florida State University Law Review* 3 (Summer 1975): 331–47.

37. J. Woodford Howard, "The Fluidity of Judicial Choice," *American Political Science Review* 62 (March 1968): 49–50.

38. Martin Shapiro, *The Supreme Court and the Administrative Agencies* (New York: Free Press, 1968), p. 43.

39. See Louis Kohlmeier, "Justice Powell: For Business, a Friend in Court," *New York Times*, March 14, 1976, Business Section, p. 5.

40. David Danelski, "The Influence of the Chief Justice in the Decisional Process of the Supreme Court," *Courts, Judges, and Politics*, edited by Murphy and Pritchett, p. 503.

41. William P. McLauchlan, "Ideology and Conflict in Supreme Court Opinion Assignment, 1946–1962," *Western Political Quarterly* 25 (March 1972): 16–27.

42. David W. Rohde, "Policy Goals, Strategic Choice and Majority Opinion Assignments in U.S. Supreme Court," *Midwest Journal of Political Science* 16 (November 1972): 667.

43. S. Sidney Ulmer, "The Use of Power in the Supreme Court: The Opinion Assignments of Earl Warren, 1953–1960," *Journal of Public Law* 19 (1970): 49–67

44. Data through the first Burger Court term from Donald D. Gregory and Stephen L. Wasby, "How to Get an Idea from Here to There: The Court and Communication Overload," *Public Affairs Bulletin* 3, no. 5 (November–December 1970); later data developed by the author.

PART FOUR | THE COURT'S IMPACT

In Part Three, we looked at the ways in which the Court handles cases. Now we turn to answer the question, "What difference do the decisions make?" to look at the effect of the cases both on the national government's legislative and executive branches and on the states and local communities. Those who do not recognize the Supreme Court as a political institution seem to expect immediate, unbegrudging, and total obedience to its decisions. There have been instances of such compliance, for example, President Truman's surrender of the steel mills the day after the Court invalidated his seizure of them, but they are rare particularly when constitutional questions are at stake. Recognition of the Court's political role and of the controversial and emotional character of many of the issues with which it deals should lead to an understanding that negative reaction to, and some noncompliance with, its rulings would be inevitable. Indeed noncompliance indicates that the Court deals with questions of substance, not abstract, sterile matters of interest only to legal scholars. One might even marvel at how much compliance there has been. If we realize that noncompliance is news while quiet compliance receives little attention and if we cut through the abundant hostile rhetoric aimed at the Court, we find that compliance does occur and that the Court does achieve results. The Court's rulings have an effect when the justices overturn actions of the states and of the other branches of the national government through judicial review. However, they occur at other times as well, when the Court engages in statutory interpretation and when it sustains statutes and thus legitimates the policies of other governmental bodies.

At times there may be "outputs with no visible outcomes and no feedback," but at other times substantial effects have been attributed to Supreme Court decisions. In some instances, people may not know much about the actual outcomes which result from the Court's rulings but respond to the Court because of effects the media have attributed to a decision.* Our tendency to make the Court a scapegoat for phenomena like increased crime rates which we can't easily explain leads at times to attributing more power to the Court than it may actually have when put to the test. The Court's decisions have been said to have produced desegregation—or retarded it; to have increased crime—or protected people's rights; to have expanded free speech—or allowed our morals to sink by facilitating the sale of indecent literature; and to have forced a president from office—while increasing the president's power by recognizing executive privilege.

Individually or cumulatively, Supreme Court decisions have had definite effects, some producing support for and compliance with the rulings, others resulting in opposition. Search warrants are now obtained where police would not have bothered to do so earlier; police now give the "*Miranda* warnings.*" On the other hand, several constitutional amendments and numerous statutes have been passed to overturn the Court's decisions. Despite such clear effects, because the Court's rulings are often only one of multiple causes of events, it is often difficult to tell whether the Court has produced social change, increased its pace, or merely served as a catalyst. Southern schools are desegregated, but it is hard to tell how much as a result of *Brown v. Board of Education* and how much as a result of the statute providing for the cutting off of federal funds to government units which discriminate. Except in the Deep South, changing public opinion, itself a result of World War II, might have led to some change without court rulings, but changed national opinion certainly made it easier for the Court to rule as it did. *Brown* itself had not produced much activity except resistance in the Deep South before Congress acted, but the statute and the desegregation guidelines

*Herbert Jacob, "Impact of Outcomes on Demand and Support," *The Measurement of Policy Impact*, edited by Thomas R. Dye (Proceedings of the Conference on the Measurement of Policy Impact, Florida State University, 1971), pp. 17–18.

issued by the Department of Health, Education, and Welfare might not have occurred without the Court's action.

Decisions on desegregation and the perhaps even more widely disobeyed school prayer rulings were highly visible and affected many people, average citizens as well as officials. Other rulings, like those on reapportionment or criminal procedure, have their basic direct effect on government officials, through structural or organizational changes, although changes in policy may ultimately affect the citizens. Still other decisions may serve to alter our federal system of government. As Charles Warren stated years ago, "To untrammeled intercourse between its parts, the American union owes its preservation and strength. Two factors have made such intercourse possible —the railroad, physically; the Supreme Court, legally."* Most decisions have an economic effect, although the visibility of that effect may differ and sometimes it may come through costs imposed, for example, by increasing the number of people eligible for welfare, at other times by costs removed, as in the 1976 invalidation of the federal minimum wage for state and local employees. If some effects are specific, others can be relatively diffuse, as when the Court affects the nation's political agenda by helping focus attention on a subject like abortion. The increase in the number of abortions obtained and the wide variety of "Right-to-Life" activities following the Court's 1973 rulings also demonstrate that a decision can have both specific and diffuse effects.

In Chapter 8 we look at some of the Court's doctrines concerning permissible actions by the legislative and executive branches and at the effect of those doctrines. The Court's treatment of Congress and its statutes and some of Congress's responses are considered first, followed by the Court's treatment of the president and its review of administrative agency actions. Chapter 9 begins with a discussion of terms such as *impact* and *compliance* and of frameworks for studying communication and impact of the Supreme Court's decisions. This discussion is followed by a detailed look at channels through which such decisions might be communicated. The impact of decisions in communities, explored through studies of the effect of criminal procedure and school prayer rulings, is followed by analysis of some of the factors which help to explain both communication and impact across different policy areas.

*Charles Warren, *The Supreme Court in United States History* (Boston: Little, Brown, 1922), vol. I, p. viii.

8 | THE SUPREME COURT AND THE OTHER BRANCHES: THE IMPACT OF JUDICIAL REVIEW

JUDICIAL REVIEW: CONGRESS AND ITS RESPONSE

Judicial Review of Congress's Acts

The greatest controversy over judicial review of national legislation came in the 1930s when the Court invalidated much of Roosevelt's New Deal program and the president responded after the 1936 election with his "Court-packing" plan. The Court's adoption of a self-restrained posture with respect to legislation regulating the economy led many to believe that judicial review was no longer frequently used at the national level. This impression stemmed in part from the greater attention received by the Court's invalidation of state legislation, particularly by the Warren Court in the area of civil liberties. Judicial review of national legislation has, however, continued since 1937, and particularly after 1950 the Court has declared invalid a number of federal statutes, some for reasons related to its invalidation of state laws, while others with no state counterparts have also been struck down.

The Warren Court voided more than twenty federal laws and the Burger Court has struck down more than a dozen others. The Warren Court acted primarily with respect to citizenship, courts-martial, internal security, and the Fifth Amendment. Constitutional prohibitions against bills of attainder and cruel and unusual punishment were used to invalidate removal of a person's U.S. citizenship for desertion from the military, remaining abroad to avoid military service, or voting in a foreign election.[1] The Court said that courts-martial could not be used to try civilians, including spouses of service personnel, civilian employees of the military, and ex-servicemen (for offenses committed while still in the service), as well as servicemen themselves for non-service-connected offenses committed outside a war zone.[2]

Restrictions on Communist party members' right to travel, to be labor union officers, and to be employed in defense plants were also struck down. So were the Internal Security Act compulsory registration provisions for forcing self-incrimination under the Smith Act.[3] The Fifth Amendment was also used to void federal gambling tax stamp, firearms, and marijuana tax laws under which compliance with registration provisions caused a person to incriminate himself under state laws or other federal laws.[4] The Lindberg Act death penalty provisions, which allowed capital punishment to be imposed by the jury but not the judge, were said in *United States v. Jackson* to interfere with the right to a jury trial. And in the welfare area, durational residence requirements in the Social Security Act were set aside in *Shapiro v. Thompson* because they interfered with the right to travel.

The Burger Court's acts of national judicial review, which often involved only part of a comprehensive statute and at times occurred through summary affirmance of lower court decisions, were scattered across a wide range of policy areas. Welfare legislation limiting those who could receive food stamps, discriminating between male and female surviving spouses as to dependents' benefits, and limiting the payment of death benefits to illegitimates were all overturned.[5] A law discriminating between male and female armed services personnel as to the requirements for receiving dependents' benefits was invalidated in *Frontiero v. Richardson*. In the area of free speech, Post Office obscenity screening methods which failed to provide adequate procedural protections were invalidated, as was a prohibition on wearing a military uniform in a play if the armed services might be discredited and a ban on demonstrations on the U.S. Capitol grounds.[6] Interference with freedom of speech was also the basis, in *Buckley v. Valeo*, for voiding Federal Election Campaign Act provisions limiting expenditures by individuals on behalf of candidates of their choice.

The Court scrutinized not only congressional statutes but other congressional actions as well. In one of its most important internal security rulings, *Watkins v. United States*, the Warren Court invalidated a contempt citation resulting from a witness's refusal to answer the House Un-American Activities Committee's questions because the relationship between the questions and the committee's investigation had not been made clear; in his opinion for the Court, the Chief Justice severely criticized both the committee and Congress, the latter for failing to control the committee's activities. The Court's setting aside of Congress's refusal to seat Representative Adam Clayton Powell (D-N.Y.) has already been noted. In that case, the Court had ruled that the Constitution's Speech and Debate Clause, which provides immunity for legislators' official activities, prevented a suit against members of Congress for their votes not to seat Powell but not suits against Congress's employees. While the justices said in the *Gravel* case in 1972 that a grand jury could not question a senator's legislative assistant about activities related to the senator's legislative work, the majority also said that the senator's arranging to have a private firm publish the Pentagon Papers was not legislative activity; thus the senator's aide, and probably the senator himself, could be questioned about that subject. The *Gravel* decision and the Court's holding in *United States v. Brewster* that acceptance of a bribe—by Senator Daniel Brewster (D-Md.)—was not part of a legislator's official duties and thus could properly be the basis for an indictment not only limited the meaning of "legislative activity" but also restricted the effect of *United States v. Johnson*, a 1965 ruling that a congressman's speech on the floor of the House could not be used against him in a bribery case. Going even further, the Court then held in *Doe v. McMillan* that circulation outside Congress of a House committee report which named specific school children in connection with an investigation of the District of Columbia schools was not part of the legislative process and could be the basis for an injunction.

Congressional Response

Congress's responses to Supreme Court actions affecting its work have been varied. At times no action has been taken even when the Court's rulings would allow such action, perhaps because the forces in Congress were sufficiently evenly balanced that no agreement could be reached on a new statute, perhaps because of lack of concern. The *Toth* decisions preventing courts-martial of ex-servicemen for in-service offenses did not preclude a statute allowing them to be tried in federal district court, but such a law was never enacted, allowing some of those involved in the My Lai massacre to remain beyond the government's reach because their involvement was not discovered until after they were discharged from the service. Another example was provided by *Branzburg v. Hayes*, in which the Court said that although the First Amendment directly did not provide protection for news media personnel refusing to reveal confidential sources to a grand jury, federal and state "shield laws" were permissible. Although legislation to provide such protection was introduced in Congress, none has yet been enacted.

Negative reaction has shown up in criticism of the Court's rulings, attempts to amend the Constitution to overrule the Court, efforts to limit the Court's appellate jurisdiction, and refusal to provide compensation which the Court's rulings seemed to require. It is also evident in reluctance or refusal to provide additional judges for the federal court system or to increase the justices' salaries, in attempts to impeach justices—as in the unsuccessful effort to remove Justice Douglas—and in resistance to confirming presidential nominees to the Court. However, most of the Court's actions have not been attacked and most of the attacks have been unsuccessful. The Eleventh, Sixteenth, and Twenty-Sixth Amendments, as well as the post-Civil War amendments on slavery and the status of blacks, were passed to override Court decisions. Many more such efforts—including massive attempts on reapportionment and school prayers and a more recent attempt concerning school busing—have failed. Current efforts to amend the Constitution either to prohibit abortion completely or to allow the state to do so, despite the attention the subject received during the 1976 presidential election campaign, seem unlikely to succeed. However, indicative of the range of legislative reaction is that the Congress has imposed a statutory ban on making available funds under the Foreign Assistance Act for abortion as a method of family planning; banned Medicaid payments for abortions; banned fetal research; and prohibited Legal Services attorneys from handling abortion-related cases; and has as well, in the "conscience clause" amendment to the Hill-Burton Act, declared that no federal funds shall be denied to public or private hospitals which refuse on the basis of religious or moral beliefs to perform abortions or sterilizations.

Attempts to limit the Court's jurisdiction have been less successful than efforts to amend the Constitution. The only such provision to be enacted was the post-Civil War removal of jurisdiction under the Habeas Corpus Act of 1867. However, the Jenner-Butler bills to limit the Court's jurisdiction in five internal security policy areas failed by only one vote to win Senate passage, and the Court did retreat from the rulings which had come under fire.[7] There have been fewer broad-gauge congressional attacks on the Court as time has passed; more action has been directed to particular rulings or sets of rulings. Conflict has been more likely to arise on matters where "constitutional language is unclear and . . . on which public sentiment has been largely unsettled"; where interest groups see vital interests at stake; and where the Court has threatened Congress's authority.[8] Action reversing the Court has come more frequently in the economic sphere than in the civil liberties domain. Economic regulation issues were involved in four of the seven periods in which proposals to curb the Court were frequent, but civil liberties issues were involved in only two such periods, according to Nagel, who also found that the presence of intensely held economic or civil libertarian interests meant less success for Court-curbing than occurred in the areas of separation of powers or federalism, where Congress could take into account policy factors outside the Court's concern.[9]

Attempts to rewrite or reenact statutes have been more frequent and far more successful, although they, too, occur with respect to only a small proportion of the Court's rulings.[10] Between 1944 and 1960, the Court's actions were revised fifty times. Among the decisions altered were thirty-four instances in which the Court has overturned sixty statutes.[11] For example, protection for longshoremen working on vessels and gangplanks between the vessels and piers came in the Longshoremen's and Harbor Workers' Compensation Act (1927) to counter action by the Court, and antitrust immunity for nonsigner arrangements in "fair trade" agreements was provided by the McGuire Fair Trade Enabling Act after the 1951 *Schwegmann Brothers* ruling that there was no such immunity. Perhaps more significant, protection for workers' collective bargaining rights was enacted in the National Labor Relations (Wagner) Act after the Court voided the National Industrial Recovery Act (NIRA) in the *Schechter Poultry* case.

After judicial alteration, Congress has also returned the law to its original state. The Court's 1908 ruling that the Interstate Commerce Commission could not compel testimony to secure information in order to recommend legislation to Congress was reversed and the Supreme Court capitulated. The holding in *Wong Yang Sun v. McGrath* that the Immigration and Naturalization Service's deportation proceedings should be subject to the Administrative Procedure Act of 1946 produced a prompt statutory exemption of such proceedings from APA coverage. When the Supreme Court held in the *Wunderlich* case that judicial review after the "finality clause" of a government contract could only take place based on fraud by the government contracting officer, Congress responded by providing that government contracts must allow an appeal by the contractor to the court of appeals. This not only reinstated the *status quo ante* after the Court had disturbed a "pre-existing 'common understanding'," true in other instances as well, but provided a broader basis for judicial review than had existed before the Court's action.[12] In another action which left intact the basic thrust of what the court had done, after the ruling in *Jencks v. United States* that the government must for the purpose of testing a witness's credibility make available to the defense all records used by a witness, Congress in the Jencks Act said that only such records as a judge had determined were relevant to the witness's testimony had to be produced.

Some congressional actions, including those in which statutes have been reenacted with clarified provisions, come because the legislators feel the Court has improperly interpreted the relevant statute. For example, when the Court broadly defined coverage under the Fair Labor Standards (minimum wage) Act, Congress amended the statute to make clear it wished a narrower meaning; when the Court defined "employee" under the National Labor Relations Act to include foremen, Congress reversed that interpretation when it passed the Taft-Hartley Act. Knowing how narrowly the Court had interpreted "commerce" under the Interstate Commerce Act, Congress purposely used broad language in writing the Federal Trade Commission Act; ironically, in the 1970s the Court said that "in commerce" was not as broad as the full reach of Congress's power under the Constitution.

Other examples of congressional reaction to the Court's rulings appear in the area of federal-state relations. Thus, when the Court has said the federal government "preempted" an area, preventing state regulation, Congress has allowed the states to take the action. For example, when the Court said in *Guss v. Utah Labor Relations Board* that a state could not regulate interstate labor disputes even in the absence of National Labor Relations Board action, Congress rewrote the law to allow such state activity. When the Court held that the offshore oil lands belonged to the national government, Congress passed "quitclaim" legislation giving the lands to the states;[13] when insurance was held in the *South-Eastern Underwriters* case to be part of interstate commerce and thus subject to national regulation, Congress quickly said that state regulations should stay in force. In both these instances, the Court sustained Congress's reversal action, explicitly acknowledging Congress's intent to overturn the Court's first rulings.

These instances suggest that Congress has the last word, but other instances of interaction between Court and Congress suggest greater effect from the Court's actions. There may be a dialogue between Court and Congress when the Court, by declaring a statute "void for vagueness," indicates that the law should be written more precisely. In taking such action, the Court may even explicitly suggest what Congress must do to make a statute valid. Overturning a law regulating grain futures transactions, Justice Taft said that Congress could not regulate the transactions unless it saw them as directly interfering with interstate commerce. Congress then placed such an explicit declaration in the Grain Futures Act, which the Court upheld. Even more indicative of the Court's influence are statutes into which Congress has written language the Court has already approved. In enacting the surveillance provisions of the Omnibus Crime Control and Safe Streets Act of 1968, Congress used procedures for wiretapping under a warrant spelled out by the Court in *Berger v. New York*.

This use by the Congress of developed judicial doctrine as well as congressional reversal of the Court points to the question of whether and to what degree Congress should defer to the Court's judgment. Members of Congress, in addition to using the Court's legitimacy to support particular positions they espouse, may find it useful to rely on the Court to resolve difficult problems. They seem divided between a "judicial monopoly" position, according to which questions of constitutionality should be referred to the courts for decision by experts (judges), and a "tripartite" position in which they feel that Congress itself should decide matters of constitutionality. On questions about which the Court has not ruled, believers in the "judicial monopoly" theory see little need to think in terms of constitutionality. Those favoring referral of constitutional questions to the courts have been more likely to think that those issues were raised in the legislature primarily as political maneuvers; on the other hand, those who have wanted Congress to make its own decisions thought the questions were seriously raised in the legislature. The more junior members of Congress and those from the Middle Atlantic and Midwest were more likely to support the "judicial monopoly" approach, while those of greater seniority and those from the South and Southwest thought the courts should defer to Congress's constitutional interpretations.[14]

Congressional reliance on the Court also varies from one policy question to another. When an issue is particularly complex, Congress may stop when it has made a general policy statement—also a result of legislative compromise—and thus leave it up to the justices to make policy as the Court passes on applications of the law. Congress has been more likely to leave matters of judicial procedure up to the courts, which participate directly in rule development, although, as noted earlier, in the last few years Congress has been reluctant to approve those rules without its own examination.

THE COURTS AND THE PRESIDENCY

The relationship between Court and Congress has often been indirect, the Court ruling on statutes only after the executive branch has applied them. Most interaction between the president and the Court has also been indirect, primarily through challenges to actions by executive departments or to the president's policies after Congress has enacted them, but direct disagreement between the president and the Supreme Court has occurred at times.[15] There have been famous confrontations such as President Lincoln's rejection of Chief Justice Taney's order to release a prisoner imprisoned by Lincoln after his suspension of habeas corpus, but they have been few. Even President Andrew Jackson's famous statement, "Mr. Justice Marshall has made his decision; now let him enforce it," seemingly the epitome of resistance, involved not direct defiance but reluctance to assist in enforcing a Supreme Court mandate directed at the State of Georgia. Only five presidents (Jefferson, Jackson, Lincoln, Franklin Roosevelt, and Nixon) have been engaged in direct conflict with the judiciary, and the administration of only one (Nixon) was frequently involved in litigation.

Appointment Power and Delegation of Authority

Franklin Roosevelt's administration was the first in this century in which the Court dealt with important questions of presidential power. On questions of the president's right to dismiss employees and Congress's delegation of legislative authority to the president, the Court's actions were adverse to the president. Relying on *Myers v. United States*, in which the Supreme Court, speaking through Chief Justice (and former President) William Howard Taft, had upheld the president's authority to discharge a postmaster, FDR tried to assert his authority over the members of the regulatory commissions. When he tried to remove Federal Trade Commissioner William Humphrey and Humphrey refused to leave, Roosevelt fired him anyhow. In *Humphrey's Executor v. United States*, the Court ruled that the Congress could limit the president's power to discharge members of regulatory commissions because of political disagreement, as the commissioners were expected to play a quasi-judicial role which would require their independence.

President Eisenhower produced the same response when he removed a member of the War Claims Commission to replace him with someone of his own choosing. Even though the statute establishing the commission did not speak of removal, the Court, in *Weiner v. United States*, relying on the spirit of *Humphrey's Executor*, ruled against the president. That the Court would also impose limits on Congress's mechanisms for the choice of officials was made clear when the justices said in *Buckley v. Valeo* that the Federal Election Commission had been improperly selected because four of its six voting members were chosen by congressional leaders without presidential involvement. Because of the commission's enforcement and administrative powers, commission members were "officers of the United States" and had to be appointed by the president with the advice and consent of the Senate. Although the president could not insist on the right to remove members of the commission at will, he could not be excluded from the selection process.

In the 1930s, Congress turned over considerable policy-making authority to those administering the law. The issue of whether Congress could do so without specifying standards to guide the president's actions reached the Court during Roosevelt's first term. In the *Panama Refining* case, with liberals and conservative justices agreeing, the Court overturned Congress's grant of authority to embargo shipments of oil produced in excess of state quotas ("hot oil") because Congress had not provided standards by which the president could determine when to act. "This is delegation run riot," said Justice Cardozo. The Court also ruled in the *Schechter Poultry Corporation* case that the National Industrial Recovery Act, under which "codes of fair competition" were developed by industry groups and promulgated by the president, was infected by improper delegation. Recognizing Congress's need to delegate certain tasks to the executive branch, the Court said, however, that "Congress cannot delegate legislative power to the President to exercise an unfettered discretion to make whatever laws he thinks may be needed or advisable for the rehabilitation and expansion of trade or industry." Because the NIRA was at the heart of Roosevelt's economic program, the ruling was even more important than *Panama Refining*, but as his attention had shifted to other programs by the time of the ruling, he may not have been particularly distressed by the Court's action.

At almost the same time, the Court upheld delegation without standards in the foreign affairs area. In the *Curtiss-Wright Export Corporation* case, the justices were faced with a challenge to another presidential embargo (on the sale of arms to warring Latin American countries) imposed pursuant to a congressional grant. Justice Sutherland's opinion for the Court sustained the president's action on the basis of both executive independence in foreign policy under the Constitution and pragmatic considerations:

> In this vast external realm, with its important, complicated, delicate or manifest problems, the President alone has the power to speak or listen as a representative of the nation. He *makes* treaties with the advice and consent of the Senate; but he alone negotiates. Into the field of negotiation the Senate cannot intrude; and Congress itself is powerless to invade it. . . . The very delicate, plenary, and exclusive power of the President as the sole organ of the federal government in the field of international relations . . . does not require as a basis for its exercise an act of Congress. If, in the maintenance of our international relations, embarrassment—perhaps serious embarrassment—is to be avoided and success for our aims achieved, congressional legislation which is to be made effective through negotiations and inquiry within the international field must often afford to the President a degree of discretion and freedom from statutory restriction which would not be admissible were domestic affairs alone involved.

Curtiss-Wright is still the law, but *Panama Refining* and *Schechter* are not; in fact, those two cases, for all their strong language, were the Court's only two invalidations of delegation of authority. The Court has repeatedly accepted delegations of authority couched in terms at least as broad/vague as those involved in *Panama*, for example, in *Yakus v. United States* approving the delegation of authority to the administrator of the Office of Price Administration (OPA), under the Emergency Price Control Act, to fix "generally fair and equitable" prices which would carry out the purposes of the law. Most recently, the Customs Court ruled in 1974 that in 1971 the president had exceeded congressionally delegated authority when he imposed a 10 percent surcharge on dutiable imports, and the Court of Appeals for the District of Columbia similarly ruled that the fee increase imposed by Presidents Nixon and Ford on imported oil was outside the president's authority, although an "honest attempt . . . to find a solution to a difficult crisis." The Supreme Court, however, unanimously reversed the Court of Appeals in 1976. The justices said the statute had authorized what the president had done; standards had been provided; and the president's actions were limited so that he could do only that which was necessary to prevent damage to the national security. Moreover, the statute has also properly provided him "a measure of discretion in determining the methods to be used to adjust imports."[16]

War Powers

Unlike its rulings on the president's appointment powers and its mid-1930s rulings on delegation of authority, the Court's decisions on the president's authority as commander in chief have been favorable. Judges have generally been quite unwilling to challenge the president's war-making authority, at least until after a war is over. As Rossiter observed, "Whatever limits the Court has set upon the employment of the war powers have been largely theoretical, rarely practical."[17] For example, although *Ex parte Milligan*, invalidating the practice of trying civilians at courts-martial when the civilian courts were operating, contained stern words applying the Constitution to the President (Lincoln), it came after the Civil War had ended. Moreover, in *The Prize Cases*, involving the blockade of the South, the Court sustained Lincoln's ability to wage war without a congressional declaration of war, and later, in *Texas v. White*, it upheld his theory of the relation between the seceded states and the union. President Franklin Roosevelt's relocation of the Japanese-Americans was held valid in the *Korematsu* case despite the dissenters' claims that the relocation was racist. In fact, the Court sustained the action not only on limited grounds of "military necessity" (the idea that the Court did not have jurisdiction to review military judgments) but also specifically held the action constitutional. (The Court did, however, say in *Ex parte Endo* that, once found to be loyal, a relocated citizen must be released.) Certainly such rulings would leave the impression that "as in the past, so in the future, President and Congress will fight our wars with little or no thought about a reckoning with the Supreme Court."[18]

President Truman's seizure of the steel mills was founded in part on the war power; the government's claim was that if the mills closed, our national security would be endangered. In response to the challenge of the seizure, the government also argued the president's "inherent power" to make the seizure. Insistence on such authority not grounded in specific statute or constitutional provision led the Supreme Court to invalidate the seizure in *Youngstown Sheet & Tube Co. v. Sawyer*, although Truman's three appointees did support the president's position. As noted earlier, the mills were returned *the next day*. In the longer run, however, the case may have "taught the Chief Executive a lesson ... to avoid invoking such extraordinary powers when dealing with business and labor and to employ, instead, several lesser sanctions none of which is as potent as seizure but the cumulative impact of which enables the President to prevail."[19]

Although many efforts were made to challenge in the courts the constitutionality of our involvement in Vietnam—because Congress had not declared war or because the war was said to violate international law—judges certainly did not limit presidential authority. Using the "political question" doctrine or ruling that by making appropriations Congress had acquiesced in the war, lower courts refused to interfere.[20] Despite consistent dissents by Justice Douglas, joined occasionally by Justice Stewart and once by Justice Harlan, the Supreme Court consistently refused to grant review, even refusing to do so in *Massachusetts v. Laird*—over dissents from all three —when pursuant to state legislative action Massachusetts's attorney general sought adjudication of the war's constitutionality.

When the challenge to the bombing of Cambodia initiated by Representative Elizabeth Holtzman (D-N.Y.) was successful in the lower courts, the Court also voided ruling on the issue. The trial judge had said that because courts were often called on to decide when wars begin and end, the issue was not a "political question." He had then held there was no congressional authority to order forces to combat in Cambodia or release bombs over that nation, and issued an injunction against further bombing of Cambodia. Although Congress had passed a specific ban on further Cambodian activity to take effect less than a month later, the government sought—and won— a stay of the judge's order from the court of appeals. Justice Marshall, the Circuit Justice, then upheld the government by ruling that the appeals court had not acted improperly in issuing the stay and that as a single justice he would exceed his authority if he were to set it aside. He did say, however, that the issue might well be justiciable and that the president could not wage war "without some form of congressional approval" except in extreme emergencies. He also added the statement that "the decision to send American troops 'to distant lands to die of foreign fevers and foreign shot and shell' [Justice Black's language in the Pentagon Papers case] . . . may ultimately be adjudged to have not only been unwise but also unlawful." The persistent plaintiffs, however, succeeded in getting Justice Douglas to vacate the appeals court stay; Douglas said the case was like any capital punishment case in which one wanted to avoid having someone die unnecessarily. Not only did the Defense Department threaten noncompliance with Douglas's order but Douglas was overruled *the very same day*. Justice Marshall, indicating he had communicated with all other members of the Supreme Court, directly stayed the district court injunction, leaving as its result the Court's usual position of noninterference as to war matters.[21] As Schubert has suggested, the evidence does not suggest that the rare decision in which a majority "of Supreme Court justices shout 'Check!' at the President has had the presumably salutary effect of keeping the Presidency in rein, to say nothing of rendering it mated. The most effective restraints upon both the presidency and the Congress have been those imposed by other components of the national political system."[22]

The Nixon Administration

Judicial challenges to actions of the president had been relatively infrequently until the Nixon administration, when the courts regularly were asked to invalidate the president's acts or to force him to do what he had not done. Many of these controversies never reached the Supreme Court because the administration did not appeal adverse lower court decisions, hoping thus to limit their legal effect. Among these cases was a ruling that the president had unconstitutionally failed to submit to the Senate the name of his nominee to be director of the Office of Economic Opportunity (OEO); invalidation of his attempt to terminate the Community Action Program (CAP) element of the War on Poverty because he had not given Congress the statutorily required opportunity to act on a reorganization plan; and the overturning of the president's pocket veto of a bill (the Family Practice of Medicine Act) during a five-day Christmas recess, because Congress had ample opportunity to consider his objections when it returned from the recess.[23]

When the president impounded funds either allotted or appropriated by Congress, that issue did reach the Supreme Court. The president had said that priorities must be given to other programs or that because the expenditures would be inflationary, he would exercise restraint if Congress would not. A large number of cases were generated, each over a specific withholding of funds, and the government lost over *thirty* decisions in the lower courts involving funds for state and local education programs, community mental health centers, highway funds, and a variety of environmental programs. The Supreme Court denied an attempt by the State of Georgia (*Georgia v. Nixon*) to bring a complaint against impoundment on behalf of its citizens directly in the Court's original jurisdiction. However, the Court unanimously ruled against the president in early 1975 in *Train v. City of New York* and *Train v. Campaign Clear Water*, involving $9 billion for waste water treatment plants. On the basis of statutory interpretation, the Court said that President Nixon had exceeded his authority in refusing to allot 55 percent of the 1972 Water Pollution Control Act funding. Justice White said that Congress had clearly intended to "provide a firm commitment of substantial sums within the relatively limited period of time in an effort to achieve an early solution of what was deemed an urgent problem" and would not have intended to undercut its own program by giving the president unlimited impoundment power. Although the president lost, the decision in a way was anticlimactic, because in an indication of partial compliance with some of the earlier court rulings he had released $1 billion of impounded funds in December 1973 and Congress had passed and the president had signed the Impoundment Control Act of 1974, which provided for impoundment or deferral of funds only after congressional examination.

Far more serious than impoundment were the continuous confrontations over the release of tapes and other evidence. The president resisted subpoenas centered on various aspects of "Watergate," including requests for material for the grand jury, for the "cover-up" trial, for the trial of those alleged to be involved in the burglary of the office of Daniel Ellsberg's psychiatrist, and, on the congressional front, for the Senate Watergate Committee and the House Judiciary Committee. The president was successful only with respect to the Senate committee, defeating its request first on technical jurisdictional grounds and later escaping having to turn over material because the judge ruled that impeachment and trial processes were more important than the Senate's need for the material to draft remedial legislation.[24] The House committee chose not to litigate the president's refusal, as it could have by issuing a contempt citation against him, but turned instead to impeachment proceedings.

The president's claims were overcome by the need for information for criminal proceedings. The initial court test, not carried to the Supreme Court, involved a subpoena for evidence (tapes) for the Watergate grand jury. Judge John Sirica balanced the idea that privileges against disclosures should be as narrow as possible against the right to presidential privacy. He recognized executive privilege as a valid concept but insisted that a judge, not the executive acting by himself, must make the decision as to the validity of the president's claim. He then ordered the tapes produced for his in-chambers (*in camera*) inspection so that he could turn over only necessary portions to the grand jury, thus protecting other material. When they received the case, the judges of the Court of Appeals for the District of Columbia tried to avoid a constitutional confrontation. They first suggested that inspection of the tapes be carried out by the president or his delegate, his counsel, and the special prosecutor. Such efforts at compromise were, however, unsuccessful, and the judges then clearly rejected the president's contentions that executive privilege was absolute, that he was immune from compulsory process by the courts, and that Judge Sirica's order threatened the "continued existence of the presidency as a functioning institution." Sirica's order was upheld while being modified slightly: the president could object to turning over specific segments on particular grounds such as national security, and these particularized claims would be judged prior to the more general *in camera* inspection.[25] (Under this procedure, some claims, for example, that material was "unrelated to Watergate matters," were sustained by Judge Sirica.)

President Nixon at first acted as if he would appeal the court of appeals ruling, and he was quoted as saying he would comply with only a "definitive ruling" from the Supreme Court; he later indicated that the vote might have to be at least 7–2 before he would comply. Perhaps because it was made clear to him by those close to the Supreme Court that he would lose by at least 7–2, he then decided not to appeal; apparently Justice Rehnquist and possibly the Chief Justice would have been the only votes for his position.[26] The president then engaged in an abortive attempt to furnish a summary of relevant portions of the tape, followed by the "Saturday night massacre" of Cox, Richardson, and Ruckelshaus, and then by presidential compliance with Judge Sirica's order. Yet even the firing of Cox was found illegal, Justice Gesell ruling in *Nader v. Bork* that Justice Department regulations "having the force of law" had been violated.

The president's next refusal to comply came when Judge Sirica ordered production of a much larger number of tapes for the Watergate cover-up trial. Special Prosecutor Leon Jaworski took this issue directly to the Supreme Court, which extended its term and heard the case on an accelerated basis. Jaworski argued that the Constitution did not speak of executive privilege and that the president could not conceal evidence necessary for a criminal trial. The president's lawyer, reflecting the position that the chief executive had to reach his own decisions on constitutional matters, told the justices the matter was being submitted for the Court's "guidance and judgment with respect to the law," that there was no way to reach the president except through impeachment, and that the president himself defines the scope of his own powers and should decide whether or not to withhold evidence.

In *Nixon v. United States*, a unanimous Court, speaking through Chief Justice Burger, rejected the president's arguments and ruled that the special prosecutor had made a showing sufficient to justify the subpoena. (It did, however, decline to rule on whether the grand jury had acted properly in naming the president as an unindicted co-conspirator in the cover-up.) The justices, while asserting that in a case involving the president "appellate review ... should be particularly meticulous," emphasized their power to determine the constitutionality of claims made by other branches of government, whether pursuant to express or implied constitutional provisions. Chief Justice Burger twice used Chief Justice Marshall's *Marbury v. Madison* language that it is "emphatically the province and the duty" of the Court "to say what the law is." The Court did legitimize "executive privilege" by recognizing that the president needed "complete candor and objectivity from advisers" which confidentiality of communications would assist. However, balancing that interest against the requirements of the criminal process, the Court struck the balance against the president, who could not be above the law. An undifferentiated claim of privilege not involving protection of special types of secrets "cannot prevail over the fundamental demands of due process of law in the fair administration of criminal justice" because it would interfere with the judiciary's ability properly to handle criminal cases and would "upset the constitutional balance of 'a workable govern-ment.'"

The decision, with which the president complied, was important in lead-ing to his departure from office. The recognition of executive privilege—even though its application in the context of a criminal trial was not in the president's favor—may, however, have been a more far-reaching result of the case.

After President Nixon's resignation, the issue of his pardon by President Ford did not reach the Supreme Court. A federal judge in Michigan, how-ever, did rule the pardon constitutional, saying it was within the letter and spirit of the presidential pardon power as well as a "prudent public policy judgment."[27] The Supreme Court had also said in late 1974 in *Schick v. Reed* that the president had broad discretion to treat individually each commuta-tion and pardon he chose to issue and to attach conditions, even those not mentioned in the statutes.

Litigation over the president's tapes and papers did continue for some time. The president's words were public property and he could not retrieve them, said a judge in ruling that President Nixon had "no right to prevent normal access" to the tapes, the release of which had already been sur-rounded with protections. Judge Gesell ruled that media representatives were entitled to copies of the tape portions received into evidence at the cover-up trial. Their broadcast, however, was delayed until a plan was devised which would allow access by all on an equal basis and which would not involve profit or overcommercialization.[28] Nixon's agreement with the General Services Administration (GSA) on disposition of presidential papers was set aside both by a district judge, who ruled in *Nixon v. Sampson* that the government owned most of the contested papers and by Congress in the Presidential Recordings and Materials Preservation Act. The Supreme Court sustained that legislation, which provided for screening of materials by archivists and return of purely private material to Mr. Nixon, and in-cluded safeguards against the disclosure of materials affecting confidential communications with the president. There was to be no more intrusion, the Court said, than there had been from Judge Sirica's *in camera* inspection of the tapes. The Court swept aside the expresident's barrage of objections one by one: the Act did not violate the separation of powers nor the president's privilege to have confidential communications, it did not improperly invade his privacy or association rights (as head of the political party), nor was it a bill of attainder. On this last point, the president was a legitimate "class of one" on which Congress could legislate; Congress had engaged in nonpuni-tive lawmaking to protect materials, not in punishment. Justice Stevens, more forthright, said the statute did implicitly condemn Nixon as an unrelia-ble custodian of his papers but he approved the legislation because of Nix-on's resignation from office and acceptance of a pardon. In dissent, the Chief

Justice argued vehemently that the Court was repudiating 200 years of judicial precedent and historical practice, with fundamental principles subordinated to the needs of the situation, while Justice Rehnquist thought the Court's ruling would hinder future presidents from obtaining confidential advice.[29]

Another type of rebuff suffered by the president was perhaps more typical of the Court's judicial review because it concerned not simply the powers of the president as such but also—and primarily—the executive's implementation of statutory policy. It involved policy on wiretapping and electronic surveillance, closely associated with the administration's "law 'n order" stance. What makes the Supreme Court's action of particular interest is that its rulings in other areas of search and seizure law were quite conservative. The administration's first difficulties came over the question—like that involved in the executive privilege controversy—of who (the Department of Justice or the courts) should determine whether overheard conversations were relevant to a prosecution and thus had to be revealed to the defendant. The Supreme Court, after refusing to accept the department's own determination of nonrelevancy, held in the *Alderman* case that a defendant alleging improper interception of conversations was entitled to inspect the "logs" of the conversations. Asking for a further rehearing, the administration asserted that this would damage the national security and said the Justice Department would simply stop telling the courts of the existence of foreign intelligence surveillance which the department thought irrelevant. The Court, not intimidated by the threat, refused to change its position.

Then the Court handed the administration three defeats in cases under the Omnibus Crime Control and Safe Streets Act. The law provided for electronic surveillance without a court order in national security cases, but the administration had used it in a *domestic* security case. In *United States v. U.S. District Court,* the justices unanimously said such surveillance could not be conducted without a proper warrant; the executive branch was not to be the only judge of whether surveillance should take place, so that it could be controlled by judges. The justices also said that the 1968 law did not give the president any power he did not already have and Congress had meant the law's exceptions to the warrant requirement to be the only ones. In response to the administration's argument that "internal security matters are too subtle and complex for judicial evaluation," Justice Powell stated flatly:

Courts regularly deal with the most difficult issues of our society. There is no reason to believe that federal judges will be insensitive to or uncomprehending of the issues involved. . . . If the threat is too subtle or complex for our senior law enforcement officers to convey its significance to a court, one may question whether there is probable cause for surveillance.

Then, in the *Gelbard* case, the Court held that the statutory prohibition on the use of improperly obtained wiretap evidence before grand juries justified a refusal to testify. The Court also ruled against the administration when it again failed to follow procedures specified in the statute. According to the law, surveillance orders were to be authorized by the attorney general or a "specifically designated" assistant attorney general. However, in several hundred narcotics and gambling cases wiretap evidence was obtained as a result of orders authorized by the attorney general's executive assistant. The Court threw out this evidence, saying the statutory provisions were clear and should have been followed. (The justices unanimously agreed that the initial orders were invalid, although the four Nixon appointees thought extensions of those orders were proper.) In a companion case, a five-judge majority did sustain wiretap authorizations supposed to be signed by Assistant Attorney General Will Wilson but instead signed by his deputies, saying that the misidentification was insufficient to make the wiretaps unlawful where they had been properly authorized. However, Justice White added, "We also deem it appropriate to suggest that strict adherence by the government to the provisions of Title III would . . . be more in keeping with the responsibilities Congress has imposed upon it."[30]

THE COURTS AND THE REGULATORY AGENCIES

The regulatory commissions handle many more cases than do the federal trial courts; if we were to include regulatory tasks undertaken by executive branch departments, the disparity would be even greater. The courts were often hostile to the commissioners in the commissions' early years, although the courts were later to allow the agencies considerable discretion. The Supreme Court's earliest response to some of the agencies was to restrict both their jurisdiction and the weight judges were to give the agencies' determinations. Decisions like the one that the Interstate Commerce Commission could only determine whether railroad-proposed rates were reasonable led the ICC to concede that "by virtue of judicial decisions, it has ceased to be a body for the regulation of interstate carriers."[31] Similarly, the ruling (in *Gratz v. United States*) that methods of unfair competition unknown before passage of the Federal Trade Commission Act were not within the FTC's jurisdiction and that the courts would identify those methods led the commission to stop trying to prohibit new unfair trade practices. However, when the Court held in the *Winstead Hosiery* case that the FTC could deal with false advertising and misbranding, the commission's level of activity on those subjects increased. The Court's later limitation of the commission through the decision that the FTC could deal with only that false advertising which was both unfair *and* a method of competition had to be overruled by Congress. Only in 1934 in the *Keppel Brothers* case did the Court say that it would give weight to the FTC's determination of what practices would be considered unfair. The agencies were also restricted procedurally by the Court. While the Court finally came around to giving administrators greater freedom of action, the Court's restrictive rulings caused reexamination of the agencies' procedures and helped lead to their codification in the Administrative Procedure Act of 1946.

Doctrines of Deference

The courts' more deferential approach to agency action resulted both from judges' greater familiarity with the agencies, new judges' more favorable attitudes toward regulation, and a greater recognition of the necessity of letting the agencies operate with less supervision lest the courts be swamped with cases appealed from the agencies. Instead of reviewing agency action with a fine-tooth comb, that is, subjecting even the agencies' factual determinations to *de novo* review, doctrines were developed by Congress and enforced by the courts which decreased judicial oversight of the agencies' work. Judicial review was said to be unavailable in some situations; various steps had to be followed before it could be obtained; and, when a court did examine an agency ruling, its function was limited to finding "substantial evidence" to support the agency's action. This led to court affirmance of agency actions at high rates.

The courts generally prefer that agency rulings be subject to some review, but at times review is precluded by statute—and the courts follow those statutes. For example, the National Labor Relations Board's decisions as to whether to proceed with an unfair labor practice (ULP) charge may be appealed within the agency, but the decision not to proceed is unreviewable in the courts. At other times, the steps necessary to challenge the agency are such that preclusion of review is often the effect. This applies to draft board denials of Conscientious Objector (CO) status, where the only way to get into court is to refuse induction and have criminal charges be brought against you. Court review of Internal Revenue Service (IRS) rulings on such matters as revocation of a group's tax-exempt status can generally be obtained only through an action by a donor or the group for refund of taxes paid, despite injury which the formerly tax-exempt group may suffer in the meantime.

The draft board and IRS examples, which derive from statutes, are part of the general idea that those protesting administrative action should proceed through the agencies before coming to the courts. The doctrines of *primary jurisdiction* and *exhaustion of remedies* follow this idea. According to the doctrine of primary jurisdiction, even if the courts are capable of dealing with certain questions, the agencies must make determinations first because they have expert knowledge concerning the subjects they regulate. Exhaustion of remedies means that all possible channels within the agency must have been followed before the courts will review a case. This is required so that the matter may be resolved within the agency, thus lessening the courts' burden, or may come to the courts in a different posture which would make it easier for the courts to decide. With primary jurisdiction, it is understood that a case will come to court; with exhaustion of remedies, it is hoped it will be resolved before that. Reinforcing both doctrines are rulings by the courts that agency jurisdiction over some subjects is exclusive, that is, the only way to remedy a complaint is to go to the relevant agency rather than to file civil litigation for damages directly against the person or company in court. Thus, in many labor disputes, the only way to proceed is through an unfair labor practice charge brought to the National Labor Relations Board. Because agency and court jurisdictions over related aspects of a problem often seem to overlap, the Court often has to "parcel out" responsibilities among the agencies and the courts.

When cases do get to court, a battery of rules are available to assist the courts in sustaining agency actions. One rule is to give great weight to agency interpretations concerning jurisdiction, substantive statutory provisions, and the agency's own rules. Thus, in holding that the Federal Power Commission did not have jurisdiction over thermal-electric power plants but only over hydroelectric ones, the Supreme Court relied on "a longstanding, uniform construction by the agency" of the statute; this interpretation deserved "great respect," particularly when it had been first developed contemporaneously with the statute's passage and where the interpretation had not been altered by Congress.[32]

Another rule is to uphold agency action supported by the Administrative Procedure Act standard of "substantial evidence" or a closely related test like whether the agency's actions are "arbitrary and capricious." For example, in a recent decision sustaining an ICC order, Justice Rehnquist said, "We inquire into the soundness of the reasoning by which the Commission reaches its conclusions only to ascertain that the latter are rationally supported."[33] The Court does not weigh the evidence presented to the commission or judge the "wisdom" of the agency's action. However, there are limits to the Court's deference; although the Court does not demand the ultimate in clarity, whatever the agencies do, they need to explain how they have reached their conclusions. Thus in 1972 the Court, while sustaining broad jurisdiction for the FTC, invalidated one of its orders because the commission's opinion did not, "by the route suggested, [link] its findings and its conclusions." The Court sent the case back to the agency for another try at a proper opinion.[34]

The Court might limit agency discretion by requiring due process protections for those affected by agency actions, but it has not done this frequently. Perhaps the most notable application of due process came in the Court's 1970 *Goldberg v. Kelly* and *Wheeler v. Montgomery* rulings that welfare benefits could not be terminated without a prior evidentiary hearing even though a full HEW "fair hearing" need not be held until after termination; however, in *Mathews v. Eldridge* (1976) the Court refused to extend its earlier rule to termination of disability benefits. More typical are procedural rulings in which the Court refused to require full hearings for all aspects of new drug applications, allowing summary proceedings instead; said that, while the Federal Power Commission (FPC) had to consider the anticompetitive effects of a utility's securities issue, the agency did not have to hold hearings in every case; or did not require the secretary of transportation to make formal findings even though he had to follow certain procedural requirements in making his determination that there was no "feasible and prudent" alternative route for an interstate highway scheduled to go through a park.[35]

A new area in which the Court's deference to the agencies has been tested concerns their disclosure of informal policies and information which has gone into their decisions, which the agencies prefer not to disclose so that they can maintain freedom of action. Since its passage in 1965, the Freedom of Information Act (FOIA), which has a general policy of disclosure but many exemptions, has been the subject of much litigation in this area. The Supreme Court, in its first ruling under the statute, *E.P.A. v. Mink*, upheld the Environmental Protection Agency's refusal to provide information to members of Congress about an underground nuclear test because the relevant documents, classified Secret or Top Secret, were exempted from disclosure under the statute. Here the Court deferred to Congress's action, which was later changed through FOIA amendments which allowed challenges to classification of documents. While the Court has strenghtened the agencies' hands by saying that where documents were not final opinions they need not be released, it has also ruled in favor of disclosure—both where documents were considered to be "final opinions" and where agency materials were such that their disclosure would not be a "clearly unwarranted" invasion of privacy.[36]

Support of the Agencies

That the courts generally sustain the agencies does not necessarily mean that the agencies eagerly follow the courts' rulings when those rulings go against them. Indeed there is agency resistance, followed by judicial reversal of the agency. This shows that the Supreme Court is "part of a continuous decision-making process, where its decisions reflect agency reaction to congressional actions or statutes."[37] Such a picture appears most clearly when we look at particular policy areas.

In the area of patent law the Supreme Court's impact has come less from individual decisions than from their collective force as the decisions have been reinforced by rulings of the lower federal courts. During the 1930s the Supreme Court quietly changed its standards concerning the patentability of inventions by substantially increasing the rate at which it invalidated patents; it was the results rather than the Court's language which mattered. Those decisions from the 1930s and further Supreme Court rulings in 1941, 1949, and 1950 were frequently cited by both the lower courts and others interested in patent policy. However, without a "clear command" from the Court, the Patent Office chose to follow the Supreme Court's doctrine rather than its actions. The agency thus did not change its decision making even when it was criticized by the Supreme Court for departing from the Court's standards. The Patent Office's insistence on approving patents while many of the courts were invalidating them has meant that the presumption of validity frequently applied to the actions of other administrative agencies by the courts has not been applied in the patent area; roughly half the courts of appeals opinions dealing with the subject have argued against the presumption.[38]

One reason for Patent Office resistance was that, although some lower court judges recognized the Supreme Court's change in posture and shifted their positions accordingly, others, among them the Court of Customs and Patent Appeals, resisted the change. This lower court reluctance to recognize what the Supreme Court had done meant that the justices, to try to make their point, had to reverse those courts frequently. When either the appeals courts or district courts found a patent valid, the likelihood of reversal by the Supreme Court was very high. Even when a new patent statute was enacted—partly a result of the Supreme Court's actions—little appeared to change. The Supreme Court's declaration that the new law was in fact a codification of the old law reinforced the pattern of divergent interpretations: some courts of appeals applied Patent Act language, others applied a different standard, and some appeared to apply no particular standard.

Another area of administrative policy making in which agency resistance to the Supreme Court took place was that of bank mergers. Under the 1960 Bank Merger Act, the agency (the Comptroller of the Currency, the Federal Reserve Board, or the Federal Deposit Insurance Corporation) with jurisdiction over a particular bank had to give advance approval of a merger, but the Department of Justice through its Antitrust Division could still attack the merger. Indeed, the Antitrust Division won its bank merger cases in the Supreme Court from 1963 through 1972, with the Supreme Court reversing every lower court loss by the division through 1970. The Supreme Court particularly limited the bank agencies' discretion by saying that the trial courts should subject the decisions of those agencies to *de novo* review. Illustrating the point that "administrative agencies will resist Supreme Court policies which run counter to their own interests and goals," the Comptroller of the Currency, "an unabashed defender of the interests of the banks," paid little heed to the Supreme Court's rulings. When the Burger Court favored rulings, thus reversing the Court's trend, there was a definite effect on the Antitrust Division; after 1972, the number of challenges to mergers brought by the division decreased to virtually none.[39]

Another instance of Justice Department difficulty with the Court over antitrust policy and of limits to the Court's deference to the executive came when the department settled a case in which the Supreme Court had earlier ordered one company to divest itself of another. Justice Douglas, saying, "No one except this Court has authority to alter or modify our mandate," made clear that the Court was not questioning the attorney general's power to settle suits but that the power had to be exercised within the bounds established by the Court's doctrine.[40] When the department later agreed to dismissal of an appeal from a new lower court order in the same litigation, the Supreme Court also set aside the new order despite Justice Harlan's complaint that the Court had set out "to thwart the Department of Justice when it decides to terminate an antitrust litigation."[41]

The Federal Power Commission's resistance to the Court provides an example of the Court's desire to have an agency exercise jurisdiction the agency is unwilling to use—in this case, over natural gas produced by "independents." In 1947 the Supreme Court refused to exempt producers' sales to interstate pipeline companies from the commission's jurisdiction.[42] Efforts to bring about the exemption of independent producers by statute, begun earlier, continued after the Supreme Court's action, and a new law was passed in 1949 only to be vetoed by President Truman. Then, despite the Supreme Court ruling and even without the necessary statutory changes, the FPC said that it would not regulate independents. It ruled in 1951 that "production and gathering" did include the sale of gas to interstate pipelines and thus was exempt from regulation. In the *Phillips Petroleum* case, the Supreme Court reasserted its earlier position, reversing the FPC. This time the commission ordered a freeze on the wellhead price of natural gas. However, the producers, encouraged by President Eisenhower, succeeded in gaining a legislative exemption which would have specifically overruled the Supreme Court decision but, upon the discovery of a campaign contribution by the gas interests to Senator Case (R-S.Dak.), the president vetoed the legislation. When it did try to carry out the Court's requirement that it regulate independents, the FPC encountered problems. However, when the agency shifted to "area pricing" (rather than setting rates for individual procedures), the Court sustained its work.[43]

Most other subsequent judicial activity concerning the FPC has taken place in the appeals courts rather than in the Supreme Court and thus has been more typical of the pattern for other agencies. In that pattern, the agency, although trying to anticipate the courts' reaction from accumulated judicial rulings, initiates policy changes, and the courts decide whether to accept those changes intact or to require modifications. Throughout, "the typical pattern of interaction . . . has been one of adjustment and accommodation to divergent viewpoints," with cooperation rather than conflict predominating, as the courts learn from the agencies and do not simply tell them what to do.[44]

Overall, the Supreme Court's rate of affirmance of agency rulings has usually been in the vicinity of 70 to 80 percent. During the 1947–1956 Terms, average support for the agencies was 69 percent; almost the same rate applied to the 1953–1959 Terms and a higher rate (78 percent) for the 1960–1965 Terms. In the 1957–1968 Terms, there were similar high rates, and in the 1967 Term, the rate went to 84 percent.[45] Even with the Burger Court's supposed greater conservatism on matters of economic regulation, 76 percent of the agency actions challenged in the Supreme Court were sustained in the 1971–1973 Terms.[46] The figures are particularly striking because only a very small percentage of the regulatory agencies' rulings are appealed to the courts and even fewer—the "difficult" and controversial decisions—to the Supreme Court, which denies certiorari in most cases. This means that the support level for the agencies is extremely high, with very, very few of their decisions overturned.

Not all agencies have been supported at the same level, however. During the Roosevelt Court, the basic range of support ran from a high of 86 percent (NLRB) to a low of 60 percent (FTC), although the FCC won fewer than half its cases before the Court. The Federal Power Commission and the Federal Trade Commission were supported at rates of over 90 percent during 1957–1968, and the NLRB and Internal Revenue Service won about three-quarters of their cases, but the ICC was supported less than two-thirds of the time and the Immigration and Naturalization Service won only 56.3 percent of its cases.

These differences are explained largely in terms of the agencies' substantive policies. At least through the Warren Court, Schubert has argued, the Court supported the agencies because the agencies decided cases in the same liberal direction to which the justices were favorably inclined. Thus, where agency results were conservative, the Court was more likely to overturn the decisions. A comparison of the Burger Court with the late Warren Court 1965 Term shows that the Burger Court reversed a higher percentage of liberal agency actions than had the Warren Court, although it also reversed one-third of conservative agency actions. The result was that while over 75 percent of the Supreme Court's administrative agency cases during the

Warren Court term had liberal outcomes, this was true of only 57 percent in the 1971–1973 Terms.[47]

The Court's overall supportive position toward the agencies masks differences among justices. For the 1947–1956 Terms, all except Black were consistent in overall support for the agencies. However, justices' value preferences appeared to affect their voting in some situations, for example, Black and Douglas (pro-union) and Chief Justice Vinson (anti-union) in labor cases and Black and Douglas in business competition cases. Black, Douglas, and Frankfurter were more likely to oppose an agency ruling when a person's freedom was involved than when it was not. All the justices were more likely to support the agencies when evidentiary questions were involved than when they were not, but some (Jackson, Reed, Burton) were particularly likely to do so; questions of statutory authority and due process of law also affected some voting in the Court.[48] During 1957–1968, differences between justices followed the Court's overall pattern. The justices' basic substantive policy attitudes had a great effect on their voting than "due process" and "statutory authority and interpretation" dimensions of cases.[49]

Overall, the picture we find in examining Supreme Court judicial review at the national level is that the Court only infrequently reverses the actions of Congress, the president, or the regulatory commissions and executive branch agencies. However, it is not hesitant to do so and can have considerable effect when it does. Yet, despite an apparent lack of reaction to many of its decisions and more willing compliance in other instances, Congress rewrites statutes and at times takes more severe reversal action, and the regulatory agencies have shown themselves quite capable of resisting the higher court through delay and persistence. Judicial review certainly did not cease after 1937, although the focus of such actions shifted from economic regulation to other matters, most notably in areas of civil liberties policy. These actions and their aftermath show the Court important but not all-powerful. From this examination, we now turn to look at the way in which the Court's rulings are communicated to those, particularly those outside Washington, D. C., who are expected to respond to and abide by the decisions. Here we will look particularly closely at the response to some of the Court's criminal procedure decisions and to its major church-state rulings, a discussion we will follow with an examination of factors affecting both the communication and impact of the decisions.

NOTES

1. *Trop v. Dulles*, 356 U.S. 86 (1956); *Kennedy v. Mendoza-Martinez*, 372 U.S. 144 (1963); *Afroyim v. Rusk*, 387 U.S. 254 (1967).

2. *Reid v. Covert/Kinsella v. Kruger,* 354 U.S. 1 (1957); *McElroy v. Guagliardo,* 361 U.S. 281 (1960); *United States* ex rel. *Toth v. Quarles,* 350 U.S. 11 (1955); *O'Callaghan v. Parker,* 395 U.S. 258 (1969).

3. *Aptheker v. Secretary of State,* 378 U.S. 500 (1964); *United States v. Brown,* 381 U.S. 437 (1965); *Robel v. United States,* 389 U.S. 258 (1967); *Albertson v. Subversive Activities Control Board,* 382 U.S. 70 (1965).

4. *Marchetti v. United States,* 390 U.S. 39 (1968); *Haynes v. United States,* 390 U.S. 85 (1968); *Leary v. United States,* 395 U.S. 6 (1969).

5. *U.S. Department of Agriculture v. Murry,* 413 U.S. 508 (1973), and *U.S.D.A. v. Moreno,* 413 U.S. 529 (1973); *Weinberger v. Wiesenfeld,* 420 U.S. 630 (1975); *Jimenez v. Weinberger,* 417 U.S. 628 (1974).

6. *Blount v. Rizzi,* 400 U.S. 419 (1971); *Schacht v. United States,* 398 U.S. 58 (1970); *Chief of Capitol Police v. Jeannette Rankin Brigade,* 409 U.S. 972 (1973).

7. See Walter F. Murphy, *Congress and the Court* (Chicago: University of Chicago Press, 1962).

8. Stuart Nagel, *The Legal System from a Behavioral Perspective* (Homewood, Ill.: Dorsey Press, 1969), p. 275; Murphy, *Congress and the Court,* pp. 257–258.

9. Nagel, *The Legal System,* p. 266.

10. An earlier version of part of this material appeared in Stephen L. Wasby, *The Impact of the United States Supreme Court: Some Perspectives* (Homewood, Ill.: Dorsey Press, 1970); see pp. 203–13.

11. Samuel Krislov, *The Supreme Court in the Political Process* (New York: Macmillan, 1965), p. 143.

12. Note, "Congressional Reversal of Supreme Court Decisions: 1945–1957," *Harvard Law Review* 71 (May 1958): 1336. For recognition of what Congress had done after the *Wunderlich* case, see *S & E Contractors v. United States,* 406 U.S. 1 at 13–14, 25–26, 48–51, and 69–90 (1972).

13. Lucius J. Barker, "The Offshore Oil Cases," *The Third Branch of Government,* edited by C. Herman Pritchett and Alan F. Westin (New York: Harcourt, Brace, and World, 1963), pp. 234–74.

14. Donald G. Morgan, *Congress and the Constitution: A Study of Responsibility* (Cambridge, Mass.: Harvard University Press, 1966), p. 367. See pp. 10–11 for an exposition of the theories.

15. For greater detail, see Stephen L. Wasby, "The Presidency Before the Courts," *Capital University Law Review* 6 (December 1976): 35–73, in which some of this material was first presented.

16. *Federal Energy Administration v. Algonquin SNG,* 96 S.Ct. 2295 (1976).

17. Clinton Rossiter, *The Supreme Court and the Commander in Chief* (Ithaca, N.Y.: Cornell University Press, 1951), pp. 127–28.

18. Ibid., p. 131.

19. Arthur S. Miller, *The Supreme Court and American Capitalism* (New York: Free Press, 1970), p. 100. For a complete account of the case, see Alan Westin, *The Anatomy of a Constitutional Law Case* (New York: Macmillan, 1958).

20. See Anthony D'Amato and Robert O'Neil, *The Judiciary and Vietnam* (New York: St. Martin's Press, 1972).

21. *Holtzman v. Schlesinger*, 414 U.S. 1304 and 414 U.S. 1316; *Schlesinger v. Holtzman*, 414 U.S. 1321 (1973). The early lower courts rulings were *Holtzman v. Richardson*, 361 F.Supp. 544, 553 (E.D.N.Y. 1973). After the activity in the Supreme Court, the court of appeals reversed the district court order and remanded with instructions to dismiss. 484 F.2d 1307 (2nd Cir. 1973).
22. Glendon Schubert, *Judicial Policy-Making* (Glenview, Ill.: Scott, Foresman, 1965), pp. 59–60.
23. *Williams v. Phillips*, 360 F.Supp. 1363 (D.D.C. 1973), aff'd, 482 F.2d 669 (D.C. Cir. 1973); *Local 2677, American Federation of Government Employees v. Phillips*, 358 F.Supp. 60 (D.D.C. 1973); *Kennedy v. Sampson*, 364 F.Supp. 1075 (D.D.C. 1973).
24. *Senate Select Committee on Presidential Campaign Activities v. Nixon*, 366 F.Supp. 51 (D.D.C. 1973), 370 F.Supp. 521 (D.D.C. 1974).
25. *In Re Subpoena to Nixon*, 360 F.Supp. 1 (D.D.C. 1973), 487 F.2d 700 (D.C. Cir. 1973).
26. Louis Kohlmeier, "Supreme Court Insiders Say Nixon Would Have Lost Appeal, 8–1 or 7–2," *Boston Globe*, October 29, 1973, p. 15.
27. *Murphy v. Ford*, 390 F.Supp. 1372 (W.D.Mich. 1975).
28. *United States v. Mitchell*, 386 F.Supp. 639 (D.D.C. 1975).
29. *Nixon v. Administrator of General Services*, 97 S.Ct. 2777 (1977).
30. *United States v. Giordano*, 416 U.S. 505 (1974); *United States v. Chavez*, 414 U.S. 562 (1974).
31. Quoted in Merle Fainsod, Lincoln Gordon, and Joseph Palamountain, Jr., *Government and the American Economy*, 3rd ed. (New York: W. W. Norton, 1959), p. 260.
32. *Chemehuevi Tribe of Indians v. FPC*, 420 U.S. 395 (1975). For an earlier ruling on this doctrine, see *Udall v. Tallman*, 380 U.S. 1 (1965).
33. *United States v. Allegheny-Ludlum Steel Corporation*, 406 U.S. 742 at 749 (1972).
34. *F.T.C. v. Sperry & Hutchinson Co.*, 405 U.S. 233 (1972).
35. *Weinberger v. Hynson, Westcott & Dunning*, 412 U.S. 609 (1973); *Gulf State Utilities Co. v. F.P.C.*, 411 U.S. 747 (1973); *Citizens to Preserve Overton Park v. Volpe*, 401 U.S. 420 (1971).
36. *Renegotiation Board v. Grumman Aircraft*, 421 U.S. 168 (1975); *N.L.R.B. v. Sears, Roebuck*, 421 U.S. 132 (1975); *Department of the Air Force v. Rose*, 96 S. Ct. 1592 (1976). See also *F.A.A. Administrator v. Robertson*, 422 U.S. 255 (1975).
37. William O. Jenkins, Jr., "The Role of the Supreme Court in National Bank Merger Policy," paper presented to the American Political Science Association, 1975, p. 3. An extremely helpful discussion of the ability of regulatory agencies to avoid full compliance with court decisions can be found in Robert Rabin, "Lawyers for Social Change: Perspectives on Public Interest Law," *Stanford Law Review* 28 (January 1976): esp. pp. 242–52.
38. On patent law, see Martin Shapiro, *The Courts and the Administrative Agencies*, pp. 143–226; Lawrence Baum, "The Federal Courts and Patent Validity: An Analysis of the Record," *Journal of the Patent Office Society* 56 (December 1974): 758–87.
39. Jenkins, "The Role of the Supreme Court," pp. 22, 24.

40. *Cascade Natural Gas v. El Paso Natural Gas,* 386 U.S. 129 at 136 (1967); the earlier case is *United States v. El Paso Natural Gas,* 376 U.S. 651 (1964).
41. *Utah Public Service Comm. v. El Paso Natural Gas,* 395 U.S. 464 at 476 (1969).
42. *Interstate Natural Gas Co. v. F.P.C.,* 331 U.S. 682 (1947).
43. *Permian Basin Area Rate Cases,* 390 U.S. 747 (1968). Another instance in which the Court gave the FPC jurisdiction it didn't exercise came when the Court said the agency did have jurisdiction to correct discrimination between wholesale power rates (under its control) and retail rates (not within its jurisdiction) by regulating the former even when, by themselves, the former would be reasonable. *FPC v. Conway Corp.* 96 S.Ct., 1999 (1976).
44. Daniel J. Fiorino, "Judicial-Administrative Interaction in Regulatory Policy-Making: The Case of the Federal Power Commission," paper presented to the American Political Science Association, 1975, pp. 28, 30.
45. Joseph Tanenhaus, "Supreme Court Attitudes Toward Federal Administrative Agencies," *Vanderbilt Law Review* 14 (1960–61): 473–502; Bradley C. Canon and Michael Giles, "Recurring Litigants: Federal Agencies Before the Supreme Court," *Western Political Quarterly* 25 (June 1972): 183–91; Glendon Schubert, *The Constitutional Polity* (Boston, Mass: Boston University Press, 1970), pp. 37–38.
46. Data from memorandum by John Rink, Southern Illinois University at Carbondale.
47. Ibid. Interestingly, a smaller percentage of agency actions challenged in the Burger Court were liberal than had earlier been the case.
48. Tanenhaus, "Supreme Court Attitudes."
49. Canon and Giles, "Recurring Litigants," pp. 189–90.

9 | COMMUNICATION AND IMPACT

SOME THEORETICAL CONSIDERATIONS

Impact and *compliance* are terms used frequently when people talk about the effects of a Supreme Court ruling, but the two terms do not have the same meaning.[1] Impact includes all the effects resulting from a Supreme Court decision regardless of whether people knew about the decision. Compliance may mean several things. One may distinguish between impact and compliance by saying that impact stands for the consequences of a policy —including but not limited to compliance—while compliance refers to the process, which occurs prior to impact, by which individuals accept decisions. Or one can define compliance differently, saying that a person cannot comply with a law unless he knows of its existence. In this definition, compliance is obedience to a ruling *because* of that ruling, particularly when a person through either opposition or neutrality did not previously intend to take the action required by the decision. A Court ruling may have a series of effects which may indirectly induce behavior congruent with what the Court requires; in still another definition, such congruent behavior is included as part of compliance, which would then be any behavior parallel to or congruent with the Court's ruling. Such a definition of compliance, however, comes very close to making impact and compliance synonymous, thus destroying the value of having separate definitions.

While determining what is compliant is difficult, there is little question that noncompliance exists. By any standard, some resistance is obvious, for example, Virginia's plan of Massive Resistance to school desegregation and the closing of schools in Prince Edward County. Other types of resistance

215

such as attempts to pass constitutional amendments to override the school prayer and reapportionment rulings may, however, be accompanied by compliant behavior. Further complicating matters is the concept "evasion" —behavior which appears to fall between outright defiance and full acceptance—in which the Court's ruling is accepted literally and narrowly while other ways are found to achieve the goals to which the Court has posed obstacles. For examples, when states were told in 1917 that they could not prevent blacks from residing in particular neighborhoods, people developed private restrictive covenants to achieve the same goal and then successfully enforced them in state courts. Such activity would not be considered noncompliance by those who feel that, as a matter of law, a Court ruling is binding only on the immediate parties. Yet if we recognize that the Supreme Court is deciding cases not only for the litigants before it, that it is a national policy maker, and that compliance should be evaluated in terms of obedience to the spirit as well as the letter of Supreme Court decisions, evasion is in some measure noncompliance.

In part because of such conceptual disagreements, no well-developed "theory of impact" has been developed, although some attempts have been made to develop theories of why people respond to the Court and to identify factors which affect impact and compliance. Among the former are theories based on the dynamics of personal attitudes, including the theory of cognitive dissonance—the idea that through a variety of mechanisms one tries to reduce the amount of inconsistency or conflict among one's attitudes.[2] For some people, compliance may depend on agreement with the decision or the attribution of legitimacy to the Court, while for others compliance may be only instrumental, to achieve a goal, so that their attitudes may not matter. Related to these attitude-based theories are applications of utility theory. According to those advocating this approach, "a person with the capacity to either comply or not comply with a given law will not comply when the utility of noncompliance is greater than the utility of compliance (that is, of engaging in the available alternative activity expected to yield the greatest net gratification)."[3] Obviously, such a calculus is not performed by all people in all situations, or even by most people, but it does seem to explain much impact-related behavior.

Some other theories, which focus on communication, derive from the fact that someone's failure to follow a Court ruling does not necessarily mean disobedience of the Court but may mean only that the decisions have not been communicated at all or have been distorted. Thus incomplete or ineffective communication will mean lessened—or different—impact. Among the frameworks useful for understanding the communication of Court decisions are those based on organizational theory and the diffusion of innovations. The problems faced by the Supreme Court in communicating its decisions to the lower courts and to others are much like those in any formal organization because "messages which travel downward in a hierarchy frequently suffer distortion in the process of communication" as a result of both intentional and unintentional changes in the message. Although many of a superior's commands fall within subordinates' "zones of indifference" so that they comply, subordinates' own interests not only affect their perceptions of messages but also affect the way they implement them. However, subordinates also believe they should implement superiors' directives, and in a judicial system, where "authority relationships are unusually strong," pressure to adhere to higher court rulings is substantial.[4]

The diffusion of judicial innovations, defined as "(a) a rule or set of rules requiring (b) new practices which (c) is embodied in a judicial decision," are affected by some of the same factors which affect the diffusion of other innovations. Thus, "if judicial decisions are to have tangible effects they must overcome the same kinds of lapses and distortions in communications, indifference, and occasionally hostility enountered by other kinds of change."[5] The multiplicity of channels through which information about innovations might reach recipients is of particular importance. A judicial innovation which travels through more than one channel has a greater likelihood of being received correctly; redundancy reduces distortion. Also quite important is the "social structure" of the units receiving the message: some, like federal regulatory agencies which regularly monitor court decisions, receive them through several channels; others have no such method for learning about judicial rulings nor the inclination to do so.

COMMUNICATION OF DECISIONS

Several relevant "populations," identified in terms of their roles and thus varying from issue to issue, are involved in the process by which Supreme Court rulings are communicated.[6] The populations are an "interpreting population," usually a lower court, which refines—makes clearer—the higher court's policy; an "implementing population" which applies the Court's basic policy directive; a "consumer population" for whom the directive was intended and to whom it is applied; and a "secondary population." This last grouping includes the general population not included in the consumer population—for example, for school desegregation, perhaps those without children in school, as well as both governmental and nongovernmental "attentive publics"—those interested in but not directly involved in the particular policy and its implementation.

There are several types of behaviors which occur during the communication process and which follow from it. These are the "acceptance decisions," in which populations decide whether or not to comply with the policy; "subsystem adjustment," in which behavioral norms and formal rules may be changed to accommodate the new policy and mechanisms for implementation may be developed; "compliance behavior," as the implementing and consumer populations in particular act on the policy which has been transmitted to them; and finally, "feedback behavior," the transmission of reactions to policy back in the direction from which the policy came. In this context, feedback behavior would include public opinion and the bringing of new litigation to enforce or alter the Supreme Court's directives.

The most obvious means of communicating the Supreme Court's rulings are its own opinions, available in several forms, including the "slip opinions" handed out on Decision Day, the advance sheets published by private companies soon thereafter, and, in due course, bound volumes.[7] These circulate among at least some attorneys and others interested in the Court's work. The actual availability of the decisions at the local level where people might want to use them is generally unsatisfactory, particularly in the more rural states, where a set of the *United States Reports* may be found only in the few larger cities. In many counties in the United States, a set is simply not available— and the reported decisions of the courts of appeals and district courts are even less likely to be accessible. However, even in larger cities, only a few of the larger law offices—and the county law library—may have copies.

The court system itself can serve as a means of communication. The decision in a case is generally sent to a lower court for "proceedings not inconsistent with this opinion." As a result, "the formal judicial structure . . . provides an important channel through which a ruling is transmitted to those directly under obligation to act."[8] But Supreme Court decisions often are not communicated directly to others, instead slowly working their way down to implementing populations through intervening layers of federal and state courts. The relationship of the highest court and the courts below it, despite the court system's official structure, is not hierarchical and bureaucratic.[9] Indeed, "communication channels may be so poor that subordinates do not become aware that a superior has issued a directive," particularly the further away they are from the Supreme Court.[10] Thus first-instance trial judges are not likely to obtain information through vertical channels, but may learn about them "horizontally" or laterally from fellow judges at the same jurisdiction level. Because the decisions of some state courts are cited or adopted more often than those of others,[11] they can play an important role in this process to the extent they contain discussions of Supreme Court opinions.

With few judges monitoring higher court decisions, a Supreme Court ruling may come to a judge's attention only when a lawyer presenting a case cites it. Indeed, many judges, partly because of work pressures which limit their reading, intentionally wait until attorneys bring new legal doctrine to their attention. Yet if attorneys are unaware of cases and thus do not cite them, the judges may never be apprised of new rules, and trial lawyers' negative feelings about their clients do serve to limit their learning about the rulings. Where a substantial number of attorneys practice a particular specialty, they may have an informal organization for circulating information about relevant cases, and lawyers arguing appellate cases are particularly likely to know about relevant higher court decisions. (Nonlawyers generally must rely on the interest groups to which they belong to provide such information, but most interest groups do this only sporadically.[12])

The basic "interpreting population," judges of courts nearer the implementing population, can be exceptionally important in transmitting information about the Supreme Court. Even when they correctly apply the Supreme Court's rulings, they may do so narrowly. Lower court judges who did not like the *Escobedo* ruling—that a suspect being interrogated had to be allowed access to his lawyer—refused to apply *Escobedo* to anyone who did not already have a lawyer; similarly, judges who did not like the *Miranda* ruling did not require warnings to be given to those not in custody, and then defined "in custody" as narrowly as possible. Only four state supreme courts followed the letter and spirit of *Escobedo*, while thirty-seven followed the letter but violated the spirit by refusing to extend the ruling beyond its specific facts; five openly criticized the decision. On the other hand, all but two of the courts interpreting *Miranda* complied, with fourteen classified as "liberal" in their interpretations.[13]

While showing appropriate respect for the U.S. Supreme Court, judges have engaged in sarcasm or injected "organizational contumacy" into communication channels by criticizing the rulings, by stating concern for the effect of those decisions upon the safety of the public, and by challenging the factual premises underlying the rulings, at times even going beyond criticism to urge lower state courts not to extend disliked Supreme Court rulings beyond absolute necessity so as not to unsettle the state judiciary. This failure to be more positive about the cases they were applying, with enforcement based less on "genuine enthusiasm" than on "a sense of hierarchical duty" with the courts indicating they were acting because they had to, is not likely to kindle much enforcement activity by implementing populations.[14]

Despite such "organizational contumacy," seldom does criticism become organized as it did in 1958, when the Conference of Chief Justices of the States adopted resolutions criticizing the Supreme Court for its tendency to adopt the role of policy maker without proper judicial restraints and stating that the basis of the Supreme Court's decisions should be the Constitution and not what temporary majorities might deem desirable. Lower courts have, however, openly resisted Supreme Court rulings, particularly over school desegregation and other civil rights matters. This may require that cases be appealed several times from the lower courts before compliance occurs, as with the admission of Virgil Hawkins to the University of Florida

Law School or of James Meredith to the University of Mississippi. Similarly, Alabama's effort to prevent the NAACP from operating in that state brought the case to the Supreme Court on four different occasions before the matter was settled—but not before the organization had effectively been prevented from operating during the intervening time.

Despite the visibility of Southern resistance and the fact that resistant judges were unwilling to penalize those engaged in violation of people's rights, not all southern judges acted this way. A few were aggressive, active enforcers of individual rights, while others, more gradualist, while probably having personal attitudes not congruent with the rights required by the Constitution and statutes, would enforce the law if they were given enough evidence.[15] Similarly, while some state supreme court judges (States' Righters) have emphasized local needs and problems and stressed the primacy of state law and state judicial processes, others (Federals) have been willing to take their cues from higher federal courts and from the Supreme Court's interpretation of the national constitution.[16] If some local judges were to do what the Supreme Court wanted and to resist pressures from the communities to which they had ties, they needed a "hierarchy of scapegoats" as well as more specific guidance as to how to achieve desegregation.[17] In this situation, the Supreme Court's making a virtue of district court discretion in implementing desegregation was thus counterproductive. It was also easier for state judges to sustain civil rights where they could avoid mentioning the U.S. Constitution and could emphasize state legal symbols and could cite state cases in preference to federal ones.

Trial court judges are crucial in the transmission of the Supreme Court's decisions to some groups, such as the police. "To the average officer 'the law' concerning arrest, search, and other police practices is in large measure represented by his direct and indirect knowledge of the attitudes of the local judiciary."[18] Yet the trial judge seldom explains his decision, most trial court decisions are either unwritten (announced from the bench) or unpublished, and few local government units have established methods for systematically gathering and transmitting information from the courts. Furthermore, trial judges may not even pay much heed to the Supreme Court, particularly if they have adopted a position on an issue before the Supreme Court has spoken or if their attitudes are "unfavorable to the intrusion of 'law' into court proceedings which are highly routinized."[19]

Attorneys general and prosecuting attorneys are among the lawyers playing a most important role in the communication process. State attorney generals' advisory opinions, which may incorporate Supreme Court rulings, do not have the force of law but are often given great weight by officials contemplating the legal ramifications of proposed actions. However, the advisory opinions usually are issued only on request by a public official and are not widely circulated, thus limiting their use as a means of communication. More important are informational meetings on new developments in the law and bulletins published for law enforcement officials, in which Supreme Court decisions are related to state statutes and judicial rulings and thus made more relevant to local officials' immediate concerns.

Local prosecutors, while they could play a large role in the communications process, may not be knowledgeable even on the subjects of cases they have tried. For example, Wisconsin district attorneys involved in obscenity cases felt some books and magazines "cleared" by the Supreme Court were obscene and other prosecutors gave many wrong answers to factual questions about Supreme Court obscenity cases; only 20 percent were rated as having high perception of Court policy (medium perception: 42 percent, low: 20 percent).[20] However, regardless of their knowledge, even where the police would like them to do so, prosecutors rarely undertake the task of telling the police about Supreme Court decisions or about how police practices might be altered to comply with those decisions. Similarly, a regular link with local police departments through which the information could be transmitted is likely to be lacking, so that officials who obtain legal information from the prosecutor usually must do so informally. This lack of legal information from the prosecutor has led an increasing number of large police departments to hire police legal advisers—lawyers whose task includes interpretation of Supreme Court decisions, the development of teaching materials, and training within the department.[21] Small departments, which cannot afford to hire these advisers, are thus left without systematic means of acquiring appropriate legal information.

Although sources of legal information inside the legal community are quite important, there are others outside the judicial system and the legal community. The mass media serve as an important initial source of information about cases, with television being the principal—as well as the most credible—source of news for the general public and newspapers providing greater detail. In the 1960s reporters covering the Court were frequently criticized for not being well trained, and for being passive and uncritical in their reporting of the Court.[22] Since then, however, the accuracy of reporting has increased; most reporters covering the Supreme Court for major newspapers, the wire services, and the television networks either have law degrees or have spent some time studying law. This results in greater legal sophistication in coverage of the Court and one no longer finds such errors as reporting a denial of review as a full decision of the Court.

The media do not transmit Supreme Court decisions intact, few even printing portions of the most important cases. Nor could they, not only for reasons of space but because the Court's opinions are too complex in their original form to be understood by most people. The wire services have been and continue to be the principal source of information about the Court for newspapers and to some extent for television as well. A study of *Baker v.*

Carr and the School Prayer Cases in the early 1960s showed that twenty-three of twenty-five papers carried reports of the reapportionment ruling on Decision Day, with fourteen stories coming from AP, five from UPI, one from the *Herald Tribune* News Service, and only two from the papers' staff writers. The picture was roughly the same for the school prayer rulings. However, starting with the second day after the opinion, although wire service domination continued, more stories were written by staff writers.[23]

The media do not treat all decisions the same. Some are dealt with more carefully or more sympathetically than others. Some are not treated at all, being lost in the "deluge" of rulings, particularly toward the end of each term of Court. Because the rulings are not directly reproduced, much of the "richness"—including the rationale or reasoning —of the Court's opinions is lost during transmission. The media are thus part of a translation process in which changes are introduced and different elements of a decision and its context are emphasized. Each medium differs in what it emphasizes about what happens at the Court. The wire services, which cover the Court on more days than the newspapers (relatively close behind) or the television networks, appear to have the greatest capacity to handle "raw word flow." Although individual newspapers vary considerably, wire service and newspaper coverage is closely related to the Court's output, but television coverage is not. When output increases, changes in the pattern of coverage occur, with more attention paid to impact and somewhat less to legal principles; television is more affected by increase in output than are the other media. However, when output reaches a certain level, the ability of the media to expand coverage to match that output ceases and coverage loses most of its "depth," tending to summaries of individual cases.

Each medium also seems to have a different "profile" in relation to "Court time": wire services and television give relatively greater emphasis to predecision coverage; newspapers, more to postdecision coverage. Television seems to add more "contextual information" than do the other media, and to report "informational content" about the decisions least; the newspapers, on the other hand, seem to be most balanced in coverage of various elements—including not only predecision material and the decision itself, but also material on the Court as an institution and trends in the Court's decisions. The inclusion of all this other material makes it difficult for people to find out what the Court has said: they may be able to read about the decision's impact without knowing what it is that is having the impact. Yet if a decision is to be applied to a variety of circumstances, its rationale—not merely the facts and the holding—must be communicated.[24]

For specialized audiences, television is clearly inadequate as a source of needed operational information and even newspapers are likely to have insufficient material about the Court's decisions. Because of these inadequacies, specialized magazines must perform an information-communication function about judicial decisions. However, while most occupational groups' "trade journals" can carry such information and some do, it usually is not well developed and is available only sporadically. Another, related source of information about legal decisions is the specialized material prepared for use in training. An advantage of such material is that the law is related to problem situations which those who must implement the rulings will face. Such details, although quite necessary, are not likely to have been provided by the courts. Such material must be written so that it can be understood; for example, materials for police officers must be written without a "dumb cop" image but so that those with a high school education or at most a couple of years of college can understand it.

As this suggests, training serves as a major means of communicating the law (statutory materials as well as court cases) to specific occupational groupings like police officers, who tend to rank training as the most effective means of learning about the law. Training takes a variety of forms, including degree programs at two-year and four-year colleges and much shorter two- or three-day or one-week in-service programs. These education and training programs must cover many subjects, so that coverage of legal matters is likely to constitute only a very small percentage of a total program unless, as is rarely the case, it is devoted solely to the law.

Another sort of educational program potentially important in communicating the Supreme Court's decisions is the general education through which most citizens pass. Like many of the others, however, it is underdeveloped or underutilized. High school courses in civics or Problems of Democracy usually contain some mention of the Supreme Court but little systematic examination of the law. Until recently, there has been very little "law-related" or "law-focused" education at the secondary level. Such courses, even if they do not contain information about specific Supreme Court rulings, may expose students to basic information about the Court and particularly to attitudes about the Court's decisional trends—exposure which will later serve as background when a person is exposed to other material about specific rulings.

The few studies we have of how people actually find out about those Supreme Court decisions intended to affect them have been carried out in the area of criminal procedure. A study of the impact of *Miranda* in four medium-sized Wisconsin cities included an examination of the ways in which patrolmen, their superior officers, and detectives found out about the case, how they rated their sources, and the relationship between these and a department's professionalism. In the least professionalized of the departments, Green Bay, more than a third of the police officers found out initially about the decision from the newspaper, an eighth from the opinion itself, and another 10 percent from training sessions; others discovered the ruling from television and radio, magazines, superior police officers, and the state attorney general. With both initial and later sources combined, about 75 percent heard about it from a superior officer and about the same proportion read about it in a newspaper, with only slightly fewer having some television exposure to the case; somewhat over 60 percent heard of the case in training sessions. This evidence confirms that "people who have no formal connection with the judiciary may be a more important source of information than any judiciary authority."[25] A full 40 percent, however, ultimately read the opinion itself. In Racine, a more professionalized department with a captain of detectives who was relied upon for legal information, a high percentage of police officers received their first information about *Miranda* from superior officers, who were also the predominant overall source of information about the decision in the department. In the most professional department, Madison, the most common initial source was the newspaper; however, more than 90 percent of the Madison officers were exposed to the decision in conference-and-training, while roughly three-quarters heard about the decision from the newspapers, a superior officer, and the attorney general.

The greater the professionalism of the department, the larger the number of sources of information and the greater the percentage of officers who received information from training sessions. Formal law enforcement sources were stressed more in the more professionalized departments. However, none of the departments had much contact with "outside" information which might have proved helpful in understanding the decision. Professionalization did not bring increased contact with nonpolice groups, but instead tended to bring about well-developed intradepartmental communication lines. Outside groups were listened to more frequently in the more professionalized departments, but that was because the groups furnished information which reinforced professional ideology. In all four departments, conference-and-training was rated the best source of information by the most people, with opinions as to the next best source varying. Those approving and those disapproving of the decision did not differ much in selecting the best source of information. However, those approving of the decisions were more likely to have received information at training sessions than were those who disapproved. While professionalization affected the way *Miranda* was communicated and received, after the decision as before, "there was no real hierarchy through which binding directives regarding the implementation of the *Miranda* decision could flow." Thus *Miranda* "did not basically change the decentralized and often unsystematic communications processes used to inform police departments about innovation."[26]

In a subsequent study, on how small (two- and three-officer) police departments in two Wisconsin counties found out about the Court's decisions, the officers seemed more aware of particular areas of the law related to their local law enforcement problems although their levels of knowledge of the law were not high.[27] Most often mentioned as the first source of information about an opinion was the newspaper, which was also most often mentioned as a general source and as the best source; second in all three categories was radio and television. The only other sources mentioned consistently were police magazines and the attorney general's office, although the officers in these small towns lacked direct contact with that office. Conferences and training sessions, which received high ratings and attendance at which gave the officers prestige, were mentioned in connection with *Miranda*, probably because of the large number of FBI-conducted sessions on it.

Small-town police chiefs in southern Illinois and western Massachusetts were particularly likely to have found out about Supreme Court decisions from bulletins and other specialized or professional literature, with media (particularly newspapers) next; law books and the Court's opinions were mentioned in Massachusetts but barely noted in Illinois. Few chiefs mentioned training programs as a primary source of information about the Supreme Court. However, when queried, they said such programs had provided some information about the Court's decisions. Particularly in Illinois, where training did not become mandatory until 1976, the chiefs felt they had not learned enough about the Court through training. In large city departments throughout the country, three types of sources of information about Court decisions—the decisions themselves, specialized police publications, and the district attorney—predominated, with the media accounting for only about 10 percent of the responses.

The most effective means of communication were thought by the small-town chiefs to be published federal and state court decisions; state and/or local prosecutors followed close behind in the rankings, and bulletins were also frequently mentioned as helpful. The mass media and personal friends were thought to be highly ineffective; opinion about specialized police publications was divided. Written communications were seen by most as more effective than oral communications in helping the officers learn about Supreme Court rulings, although some suggested that a combination of the two, e.g., training materials discussed at a conference, was better still. The officers seemed to want something like a regular (monthly) free bulletin or newsletter which came to the individual officer and which digested cases and updated information the officer already possessed.

The small-town chiefs unanimously felt it was important for police to know about the U.S. Supreme Court's decisions. Virtually all the chiefs said their men wanted to know about Supreme Court rulings, although they made distinctions between officers who were serious about their jobs and other, less interested officers.[28] In both states, the local prosecuting attorney was seen as having responsibility for providing legal advice to the police, for example, by preparing bulletins on the law and having meetings with the police on changes in the law; in Massachusetts the attorney general was seen as sharing such responsibilities. Almost all chiefs said that advice about the law was available to them, but many said they were not provided sufficient material or access to it, although the situation was seen as more favorable in Massachusetts than in Illinois. For example, two-thirds of the Illinois chiefs did not know where published versions of Supreme Court decisions were available, indicating that they made little or no use of the decisions themselves. Massachusetts chiefs, however, had both the Supreme Court and state court decisions available to them, often in their own offices, and furthermore read the decisions although they generally felt that there was not enough time to do so. Evaluating the material, a majority of the chiefs in both states said such materials were understandable by nonlawyers, although at least some officers would need assistance if they were to understand them.

IMPACT: RESPONSE IN THE COMMUNITY

As the just-discussed studies about the communication of the Supreme Court's criminal procedure decisions suggest, there is considerable variation between communities in the ways they "hear" decisions and in how they respond to them once they have heard some version of what the Supreme Court has said. The two matters—communication and response—are intimately related. Certainly if a community does not find out about a decision, it cannot respond, and if it receives a distorted version of the Court's ruling its response may be similarly distorted. Yet even if the justices' opinions come through "loud and clear" there will still be varying degrees of compliance with them. To examine this phenomenon of variable impact of the Court's decisions, we look at what are probably the two most studied areas with respect to that subject—criminal procedure and church-state relations.

Criminal Procedure

That the Court's criminal procedure rulings have had an impact on the police—or that the police think that such an impact has occurred—is clear. For example, all small-town police chiefs in western Massachusetts and southern Illinois said the Supreme Court had affected police work, although a majority of the latter said their own small town had not been affected, with the effects occurring only in big cities. Among the effects noted were that the decisions had actually improved the quality of police work. The rulings were said to have made "old-timers" reluctant to "get involved" with cases, but newer officers, particularly if trained, were more accustomed to the new rules and were not scared by them. However, while two-thirds of the officers in each state said the Court had handed down "helpful" guidelines, in neither state were responses specific about what those guidelines were and some Illinois chiefs even said that the Court's rulings were not salient for their work.[29]

The decisions in *Mapp v. Ohio* and *Miranda v. Arizona* have received particular attention. In one study, questionnaire responses from police chiefs, prosecutors, defense attorneys, judges, and American Civil Liberties Union (ACLU) officials indicated that as a result of *Mapp*, search and seizure questions were raised more frequently at trial but that there had not been overall change in police effectiveness. Attitudes were reported to have altered as a result of the ruling, with a general movement toward agreement with statements that the same rules should exist for federal and state police, that the legality of searches should be broadened, that more flexible search warrants were needed, and that safety should be emphasized more and liberty less.[30] Increased police adherence to legality in searches between 1960 and 1963 was reported, with a high positive correlation between increased police education on search and seizure and adherence to legality.

Mapp also produced a substantial increase in police education, including the development of courses for police in police academies, adult education programs, and colleges and universities. In fact, it was among the "key factors in increasing the attempts at centralization and formalization of police training procedures."[31] In New York, retraining sessions "had to be held from the very top administrator down to each of the thousands of foot patrolmen and detectives engaged in the daily basic enforcement function," with "hundreds of thousands of manhours" devoted to the task.[32] Training improved in previously nonexclusionary rule states and the FBI increased its training of state and local officers, but the exclusionary rule did not guarantee that good police training on arrest and search would take place, with pre-*Mapp* exclusionary rule states among those doing the least. Four years after *Mapp*, training remained "very spotty in both quantity and quality."[33]

Attitudes about search and seizure in states which had an exclusionary rule prior to *Mapp* differed from those in states without the rule before 1961. Similarly, state court decisions showed that the former states were far *less* likely to adopt federal search and seizure precedents after 1961, particularly if they had already begun to develop their own lines of precedents. This led to the comment that "where a state judiciary *anticipates* a federal rule, but with variations of its own, those variations are going to prove very hard to kill."[34] However, another study showed almost no relationship between earlier adoption of the exclusionary rule and implementation of *Mapp*; states which had previously adopted it seldom invoked it or narrowly interpreted it when they did, while states with the rule previously gave it "some informal honor." State supreme courts, in providing the law enforcement community with the basic law as to what constituted an unreasonable seizure, faced some sixteen search and seizure questions. An indication of how little state search law was developed in the post-*Mapp* period is that eighteen states did not deal with more than two of those questions. For only three questions did two-thirds or more of the state courts rule evidence inadmissible; five of the questions were settled in a manner limiting *Mapp*'s application through admission of the evidence, while there was no clear pattern of decision for the remaining eight questions. Nine states responded positively to *Mapp*, fourteen were in an intermediate category, and nine were negative.[35] Explicit judicial resistance to the rule was by no means uncommon; judges often indicated that they had little use for *Mapp*.[36]

The most important potential effect of *Mapp* was on police actions—arrest and search warrants—where changes could have resulted from increases in particular types of crimes, such as guns, gambling, and narcotics, as well as from the decision. Oaks's extended analysis of the rule's effect, cited by Chief Justice Burger in his argument that the costs of the rule were too high to retain it,[37] relied primarily on data from Chicago, the District of Columbia, and Cincinnati. Oaks concluded that "the data contains little support for the proposition that the exclusionary rule discourages illegal searches and seizures, but it falls short of establishing that it does not." But in a stronger statement he argued, "As a device for directly deterring illegal searches and seizures by the police, the exclusionary rule is a failure," and went on to say that police conduct not leading to prosecutions was unlikely to be affected by the rule, with little deterrent effect on prosecution-oriented activity.[38]

Doubt about Oaks's conclusion was raised by Canon, who looked at arrest, warrant, and disposition data for the period immediately after *Mapp* and for later years as well to ascertain the decisions' long-term effects.[39] In some fourteen cities without a pre-*Mapp* exclusionary rule, few search warrants had been used, but the proportion of constitutional searches, although varying from city to city, increased after the Court's decisions. Officials in large city police departments attributed the increase in search warrants primarily to the increase in narcotics traffic, with less of the change accounted for by increased numbers of police, better training, and other judicial rulings. In roughly one-third of the cities, court decisions were not thought to have had much effect on recent increases in the numbers of search warrants, but in approximately one-fourth of the cities, judicial action was thought to be a major if not the total cause. The data did suggest an interplay between increased narcotics use and judicial rulings as causes of increased warrants. Police decisions to use rather than not use search warrants turned on the existence of the exclusionary rule, but the increased use of narcotics largely explained substantial increases in use of search warrants from the late 1960s through the early 1970s. Charges dropped by police departments and prosecutors because evidence was improperly seized did not show evidence of widespread police violation of the exclusionary rule but there was also little evidence that the rule was having a very substantial effect. It was also the case that many officers saw the exclusionary rule as creating problems, including frustration and irritation at the granting of motions to suppress evidence.

In studies focusing on *Miranda* rather than *Mapp*, results from New Haven and Washington, D. C., indicated that not all four of the warnings required by the Supreme Court's ruling were given in all situations at the station house and in some instances none were. According to a Washington, D. C., field study, an even lower rate of warning-giving occurred "on the street."[40] When they were given at the station house, the warnings did not silence all defendants because detectives effectively used a variety of elements of psychological interrogation, those informed of their rights did not understand the warnings, or they could not take advantage of them because they did not know how to contact an attorney. Yet even when station house counsel was promptly available, many suspects talked before an attorney could arrive to assist them. Nor was this simply a result of lack of formal education. Yale University graduate students interrogated by the FBI in connection with draft-card violations were generally unaware of their *Miranda* rights until law professors called the warnings to their attention; even then, many continued to participate in the interrogations.[41]

In the four medium-sized Wisconsin cities where the effect of *Miranda* was studied, police knowledge of what the case required was relatively high but the decision appeared to produce little change in the conduct of interrogations. Police department statements of rules and procedures for interrogators to present to suspects before interrogation were undermined by the interrogators' discretion and by their habit of giving the warnings but not following through on department policies of allowing counsel to be present at interrogations. In short, compliance was "formal, perfunctory or rhetorical" without behavioral changes.[42] Detectives said that *Miranda* had changed the way they had to obtain evidence, particularly by forcing them to obtain it prior to beginning interrogation, but observations suggested instead that "all the departments continued to rely first on interrogations. If that failed, either because of lack of relevant information or because of a lawyer's refusal to allow questioning, only then were alternative methods relied upon."[43]

Unlike response to the controversial *Mapp* and *Miranda* rulings, except for rural southern counties most jurisdictions complied with the directive of *Argersinger v. Hamlin*, which extended the right to counsel at trial for any situations in which a jail term would be imposed. Courts interpreting *Gideon v. Wainwright*, the 1963 ruling providing counsel at trial for indigents for felonies and major misdemeanors, had usually adopted a narrow view of its requirements and imposed a "variety of different lines ... as to where the cut-off point" came for providing counsel.[44] But a year after *Argersinger*, most places appointed counsel even in traffic cases where imprisonment was a reasonable possibility, although they did not do so for nonserious traffic cases, which usually involve only a fine.

The increase in appointment of counsel led to increased staffing of public defender officers in less than 15 percent of the counties, particularly in the larger ones, but did not produce a great increase in not-guilty pleas in misdemeanor cases, so that few more trials were required than would have occurred without the ruling. However, provision of counsel may have delayed the guilty pleas which most misdemeanor defendants intended to enter when they came to court. Defense counsel—whether assigned counsel (the type primarily used in small counties) or public defenders (used in larger locations)—now had to help "process" guilty pleas, but the rationale of the decision—that the trial process become more adversary in nature—was not carried out. A somewhat later study, however, suggested that compliance with *Argersinger* had been only "token," with judges encouraging waiver of counsel and with "no coherent development of defense systems to meet the need for quality representation" which the spirit of the case demanded. Appointed counsel—often assigned immediately before a trial—were frequently inexperienced and not well prepared to represent their clients.[45]

Church-State Relations

The Court's church-state rulings not only produced much controversy but their impact attracted much attention. This was true both of decisions concerning "released time" programs and those invalidating prayers in the public schools. When the Supreme Court in the 1948 *McCollum* case invalidated a program of religious instruction operated on school property by religious personnel, those operating other programs attempted to differentiate their programs from the *McCollum* situation. Some programs were modified, most notably those using public school property during school time. Programs were most likely to continue where enabling legislation had been in existence for a long time. Not all on-premises programs were dropped, however, and state attorneys general did little to stop them. Opponents of the programs tended to be in religious minorities and did not carry the political weight necessary to bring about change.[46]

After the 1952 *Zorach* decision sustaining religious programs held off the school grounds during school hours—the so-called "released time" programs —attendance at such denominational activities returned to pre-*McCollum* levels, but subsequent growth was small.[47] Even though released time programs provided an easy answer, little was done to eliminate continuing noncompliance with *McCollum*. Some people interpreted *Zorach* to require rather than merely to permit released time programs, but states did not enact new programs, a fact which raises further questions about the force of the Court's power to legitimate a policy. However, Justice Douglas's dictum,

"We are a religious people whose institutions presuppose a Supreme Being," was often quoted subsequently by those favoring a variety of programs supporting religion.

The school prayer decisions of 1962 and 1963—*Engel v. Vitale*, invalidating a state-composed prayer, and *Abington School District v. Schempp*, striking down recitation of the Lord's Prayer and readings from the Bible—produced far more outcry. Local and state officials played a key role in deflecting the requirements of those rulings. Where a superintendent opposed the practices, he was able to use the Supreme Court's ruling to change local practices but needed the decisions to do so. However, far more numerous were officials who wanted to avoid conflict with small-town local power structures, often because the officials had other, more important goals to achieve. They presumed that compliance was taking place and did not check for violations or turned their heads after a perfunctory statement that prayer and Bible reading were to be stopped, thus allowing teachers to continue the practices. Lawsuits were the only practical way of challenging the practices, but little litigation occurred because it was extremely expensive and often had few supporters and because, as in the post-*McCollum* period, those in the minority on this emotional issue were unwilling to subject themselves to the pressure they would have felt had they attacked local practices.[48]

The national incidence of classroom prayers as reported by teachers dropped from 60 percent at the time of the decisions to 28 percent in 1964–1965 and Bible reading decreased from 48 percent to 22 percent, so it could be said that "the average teacher did feel the impact of the Supreme Court's decisions."[49] The largest number of teachers said there was no school policy on prayer or that matters were left up to the teachers. When there was such a policy, it had an effect: only 4 percent of teachers in schools opposing the prayers said them, compared to 43 percent of those in schools where the prayers were favored and 40 percent in schools with no policy. For Bible reading, the figures were much the same. The teachers' own religious observances had an effect on their response. Those attending church more frequently were more likely to lead prayers in the classroom —both before and after the Court ruled. Conservative Protestants were more likely than liberal Protestants, Jews, or Catholics to have led prayers before the ruling, while Jews and Catholics were most likely to comply and conservative Protestants least likely. In addition, the greater the teacher's seniority, the more likely the teacher was not to change his or her practice after the ruling.

There were major regional differences in compliance. Where the practices had been most widespread, in the East and particularly the South, they were least likely to be given up, but school prayer was far more likely to be discontinued in the Midwest and West. Some areas had had long-standing prohibitions against such devotional activities; areas which had adopted the practices without explicit requirements found it easy to give them up. Those areas which retained school prayer were likely to have had constitutional or statutory requirements underlying the practices. In the East, formal requirements were eliminated, leading to compliance, but in the South, requirements were not removed and were at times supplemented, making it unlikely that the practices would be changed.[50]

One of the studies of the compliance process concerning school prayer indicates both a variety of attitudinal responses to law and evidence that a person committed to complying with a Supreme Court ruling will search for social support and for information supportive of his or her position.[51] An individual's personal values will be strengthened as he or she seeks to convince others as well, although this may result in overreaction as the individual tries to "sort things out" and to arrive at a comfortable, nondissonant position. After the Supreme Court's major school prayer decision (the *Schempp* case), attitudes of community school officials about church-state relations became more definite and more positive than they had been earlier. The Court's ruling motivated them to reduce their indecision, with the social grouping of which they were a part playing an important role in the way they reacted.

Some individuals did not change their views about religious practices and sought information which reinforced their positions. The "backlashers," committed to defiance of the Court's ruling, tried to bring religious activities into the schools whenever they could and took a negative stance toward the Court itself. On the other hand, the "vindicateds" were critical of "schoolhouse religion" both before the decision and afterward. Others who did not change their personal views complied with the decisions because they generally believed or led themselves to believe that they had no choice but to comply; these were the "nulists." There were also many who changed their views toward religion in the school, but not all did so in the same way. The "converts" came to believe in the Supreme Court ruling and were committed to enforcing it, while the "liberateds" softened their views about religion in the school and moved toward convergence between their own views and what the Court required. The "reverse liberateds," however, became more rather than less favorable to schoolhouse religion; they moved away from the Court's position, not toward it.

FACTORS AFFECTING COMMUNICATION AND IMPACT

Numerous factors affect both transmission of information about the Supreme Court's rulings through the various channels of communication and the effect the rulings will have. These factors both alter the message before it reaches either the implementing or consumer populations and affect the ways in which those populations respond to information they do receive. At times the factors are strong enough to "wash out" much if not all of a decision's intended effect.

Characteristics of the Court's ruling itself and the ruling's place in a pattern of decisions are among those factors. Although if unanimity is achieved at the cost of a "murky" opinion, the impact may be lessened, unanimity is thought to increase impact because a unanimous opinion is easier to transmit than are several opinions in a single case. Concurring and dissenting opinions not only increase the information to be transmitted but also make it easier for potential opponents of the majority's view to resist compliance. A plurality opinion makes it particularly difficult for others to determine the Court's intent. Similarly, the addition of more and more cases will by itself increase the amount of "noise" in the communication system, particularly if they are decided on the basis of the "totality of circumstances" in the case—the Burger Court's initial preference in the field of criminal procedure—rather than a broad rule such as the Warren Court tended to use.

Even when there is only a single opinion, the relative clarity or ambiguity of the opinion affects its communicability. There are few decisions which, like *Miranda*, can be reduced to four or five warnings to be put on a "Miranda card." Even when a decision is clear initially, it may become less clear as the Court later explicates or limits it. Thus, the initial clarity of *Miranda* was blurred by rulings on use of confessions obtained without warnings or resumption of questioning after it had been stopped.[52] Greater clarity is generally thought to produce greater compliance; ambiguity, more noncompliance. However, because each recipient of a communication can perceive an ambiguous decision as supporting his position, ambiguity may produce some compliance. Furthermore, ambiguity increases the Court's influence over policy development by forcing people to return to the Court to obtain elucidation of the meaning of its rulings.

Just as a large number of cases increases distortion, ambiguity may also result from gaps in a doctrine. This is inevitable because the number of cases decided by the Supreme Court is very small in relation to the total number of issues on which it is asked to rule and because judges tend to decide only the issues presented to them and to do so narrowly. Thus courts can develop a "clear policy line" on only a few issues at any one time if they can do so at all.[53] However, the Court may set off far more of a furor when it first approaches an area, even if it does so in a case with a limited factual application, than it does later with more far-reaching cases, because its initial action may be unexpected. Thus *Engel v. Vitale*, although only invalidating state-written prayers in schools, drew far more heated reaction than *Schempp*, which outlawed the far more prevalent school prayer and Bible reading.

Also relevant in this situation is that the Court was more careful in delineating its second holding—in saying that not all teaching about religion had to be excluded from the school—than it had been in its first opinion, where important qualifications appeared only in a footnote. Similarly, less negative state court reaction to *Miranda* than to *Escobedo* may have come both from a realization that *Miranda* was likely to follow and from the *Miranda* opinion's greater clarity.

If those who are expected to implement decisions are to hear about them, the decisions must be highly visible. Because at times people become aware of rulings only after they have been cited by lower courts or commented on in an attorney general's bulletin, that is, only after they have traveled through lengthy communication channels, the initial signal must be strong to overcome interference and delays. Otherwise the message becomes garbled as in the game of "Gossip." A decision's visibility is affected both by its direction (who wins) and the communicator's legitimacy. Distaste for the Supreme Court may lead people to pay less attention to its decisions. Similarly, a feeling that the Court is not ruling fairly on a subject or on the basis of adequate knowledge will interfere with the reception given its decisions. For example, the Court's competence and hence its legitimacy was questioned in the police area when the Court in *Miranda*, seemed at least to law enforcement officers, to be acting without reference to the actualities of police work. This perceived action of the Court made it increasingly likely that the personal and organizational goals of the police would differ from the goals embodied in the Court's ruling.[54] On the other hand, when the Court showed recognition of the dangerous situations in which police find themselves, as in the *Terry* "stop and frisk" case, there was less hostility to the decision in the law enforcement community.

The ongoing situation into which a Supreme Court decision is injected is another factor affecting the receipt of communications and the decision's impact. A decision handed down in the midst of a crisis is likely to receive less coverage than one announced in "normal" times, unless the decision is perceived as contributing to the crisis or producing one. If a substantial change in the law has taken place immediately prior to the Court's ruling, obtaining further change soon will be difficult, but it is also difficult to get people to pay attention to the courts when the law is well settled, making it "more or less invulnerable to radical changes."[55] Both a community's long-term history and events of the immediate past may attune potential recipients to what the Court has said; a community with few major crimes is less likely to be "up" on criminal law than one in which a murder has recently occurred or in which the police want to make a major drug "bust". A community's general belief system also affects communication of and receptivity to decisions. Officials in communities which pride themselves on being "up-to-date" or "professional" might be more likely than other communities to seek out information about court decisions and to work to bring local government action into accord with those rulings.

This suggests that another part of the postdecision situation is *who* responds and how they respond—the "follow-up" a ruling receives. If elites support a ruling, average citizens and government employees may pay greater attention to subsequent communications about it. Even the absence of criticism—merely maintaining a neutral posture—may be important. However, if dominant community interests or prominent officials immediately oppose the ruling, further communication about it may well fall on plugged if not deaf ears. This was particularly true in the South with respect to desegregation, where *Brown v. Board of Education* provoked considerable segregationist oratory from gubernatorial candidates. Over three-fourths (77 percent) of the southern governorships from 1958 to 1961 were won by militant segregationists, a figure far higher than immediately before the Court ruled.[56]

The degree to which government officials enforce or attempt to enforce Court rulings, part of the "follow-up," is critical because those rulings are not self-enforcing. Executive branch officials do not often directly attack Supreme Court decisions but they may severely damage the possibilities for compliance by refusing to take firm action to implement them—true with respect to school desegregation in the Eisenhower, Nixon, and Ford administrations. At the local level, prosecutors' inaction concerning police violations of search and seizure rules does little to encourage compliance. Prosecutors may even approach the Court's rulings negatively, talking to police (in what is called "negative advocacy") about how to "get around" a ruling. People are not encouraged to learn about or to follow the rules if there are no sanctions or only limited ones for not knowing the law, with little if any reward for following Court-established rules. Thus, if new rules are to be carried out, there must be active supervision, not just executive passivity.

Sometimes a wide variety of mechanisms, including quite stringent ones, is needed before compliance can be achieved. For example, when a school district did not negotiate a desegregation plan with the Office of Civil Rights, a suit for the Department of Justice or by private parties might have been necessary to bring about desegregation. Yet in some districts compliance was not produced even after the loss of federal money, substantially more severe than court orders. This meant that some school districts had to be threatened with the loss of their state education aid or with the loss of both federal and state money before desegregation took place. Bringing about compliance would be assisted if those engaged in the follow-up had accurate feedback about the effect of various mechanisms. However, implementers are not always objective, particularly when their agencies are committed to a specific technique, and thus some sanctions are overused while others are not used sufficiently.[57]

The structure of organizations affected by a decision also plays its part in the communication of and compliance with court decisions. If an agency is arranged hierarchically, there is at least the possibility that commands can travel directly through channels to the individual who is supposed to comply. Yet an organization may be so large that what specialists or others at the "top" of the organization learn about judicial rulings cannot be effectively transmitted to the "bottom." Where units are small, division of labor and specialization may be insufficient to have a person assigned to monitor judicial rulings. A small unit located near other units may acquire information from those units "horizontally"; however, when the small unit—like a rural police department—is geographically isolated, it may have few contacts through which to acquire information. Political isolation may be as important as geographical isolation in determining communications received, with community influence in the form of daily pressure affecting how an agency responds to a judicial ruling. Community pressures vary with the size and homogeneity of the community and are particularly severe in small, homogeneous communities where role expectations reinforce each other. And an agency operating in a supportive environment may be more resistant to outside rules which demand change than an agency which is politically vulnerable and may comply to protect its position.[58]

Community pressures are an important part of an official's "work situation."[59] That work situation also includes the people from whom he takes cues as to how to behave—his "reference group"—which is particularly important when community expectations are contradictory, thus allowing the individual more discretion as to how to act. Some government employees ("locals") take their cues primarily from their own communities, while others ("cosmopolitans," particularly professionals), such as lawyers and doctors, look to those in the same occupation. If, like police officers, individuals spend much time in the company of colleagues and feel the Supreme Court is hurting their work, they may wish to avoid becoming "sore thumbs" or deviants and thus will not attempt to follow rules adhered to by their fellow workers. Such social pressure is increased when the work subculture becomes organized into unions or comparable interest groups.

Within the work situation, attitudes and legal knowledge affect compliance. Those who might be affected by court rulings wear "perceptual blinders" both when they seek information and when they are more passive recipients. Thus ignorance of the Supreme Court's work stems not only from uneven reporting but also comes from "audience effects" which help produce "widely varying responses among diverse segments of the receiving population."[60] Among those who might be affected by rulings—whether they be police with respect to criminal procedure decisions or booksellers with respect to the meaning of "obscenity"—few seek out information about them, so that the information has to be brought to them if they are to obtain it. Only 3 percent of respondents in a multistate survey of booksellers sought legal advice about sexually explicit books they sold on more than "isolated occasions," and only a "small minority" knew about the Supreme Court's obscenity decisions. As a result, most booksellers did not seem to be affected by the judicial policy of either the state courts or the Supreme Court; those who did allow their activities to be regulated in this fashion were the most restrictive in their attitudes toward what they would sell.[61]

One attitudinal item which might seem to have a particular effect on compliance would be an individual's general commitment to "obeying the law" including the dictates of the Supreme Court. However, "law-abidingness" plays a limited role when other factors enter the picture. Some people do alter their attitudes to bring them more in line with the Supreme Court's position, but many others develop a variety of techniques which allow them to retain their policy preferences without adhering to the Supreme Court's doctrine. Among these techniques are attacks on those who are seeking to enforce the decisions (condemning the condemners); appeals to a "higher morality" (an authority higher than the Supreme Court); denial of responsibility for the present situation, perhaps coupled with a claim that others are preventing compliance from coming about; and beliefs that those the decision is intended to protect (whether they be blacks or criminal defendants) have not really been injured or are instead the *real* menace to society.[62]

Despite all the obstacles both to communication of the Supreme Court's rulings and to compliance with those rulings which have been enumerated and discussed in this chapter, compliance with the Court's decisions does occur and those decisions have substantial impact. If that were not the case, we would hear far less objection to them. More important, without them, racial and sexual equality, freedom of speech and religion, and defendants' rights, as well as economic regulation, would not take the form in which they now exist. The resistance to implementation of those decisions serves all the more to remind us that the Supreme Court is not merely a "finder" of the law but an active policy maker.

Despite the views which many people have of the Court which do not cast it in such a role, and despite the public's general lack of awareness of what the Court has done—and even of how it operates—the justices maintain a commanding presence. Other officials must at least take them into account before they act; nor has the Court, even when publicly adopting a general posture of "self-restraint," been hesitant to strike down the actions both of its coordinate branches at the national level of government and of the state governments as well. In perhaps the most important test, attacks on the Court's major rulings have only very infrequently been successful, although the Court has at times backed off from some of its strongest positions. It is doubtful that we have "judicial supremacy" as the Supreme Court's harshest critics have argued from time to time, but we do have in the Supreme Court of the United States a body of individuals who, operating within the constraints of the nation's legal system, through both the full-dress treatment given some cases and the variety of other, less visible actions they take, regularly make policy for that system and the larger political and social system.

NOTES

1. Discussion of terminology can be found in Stephen L. Wasby, *The Impact of the United States Supreme Court* (Homewood, Ill.: Dorsey Press, 1970); in several of the articles in *Compliance and the Law* edited by Samuel Krislov and others (Beverly Hills, Calif.: Sage Publications, 1972); and in Robert V. Stover and Don W. Brown, "Understanding Compliance and Non-compliance with Law: The Contributions of Utility Theory," *Social Science Quarterly* 56 (December 1975): 363–75.

2. William K. Muir, *Prayer in the Public Schools: Law and Attitude Change* (Chicago: University of Chicago Press, 1967).

3. Stover and Brown, "Understanding Compliance and Non-compliance with Law," pp. 369–70. See also Charles Bullock III and Harrell R. Rodgers, *Law and Social Change* (New York: McGraw-Hill, 1972), chap. 8, "Law and Social Change: A Cost-Benefit Interpretation," pp. 181–209.

4. Lawrence Baum, "Implementation of Judicial Decisions: An Organizational Analysis," *American Politics Quarterly* 4 (January 1976): 94, 101.

5. Daniel J. Fiorino, "The Diffusion of Judicial Innovations: Court Decisions as a Source of Change," paper presented to the International Communication Association, 1976, pp. 3, 18.

6. Charles A. Johnson, "The Implementation and Impact of Judicial Policies: A Heiristic Model," in *Public Law and Public Policy*, edited by John A. Gardiner (New York: Praeger, 1977), pp. 107–26.

7. An earlier version of this section appeared in Stephen L. Wasby, *Small Town Police and the Supreme Court: Hearing the Word* (Lexington, Mass.: Lexington Books, 1976).

8. Richard Johnson, *The Dynamics of Compliance* (Evanston, Ill.: Northwestern University Press, 1967), p. 61.

9. Walter F. Murphy, "Lower Court Checks on Supreme Court," *American Political Science Review* 53 (December 1959): 1017–31.

10. Baum, "Implementation of Judicial Decisions," p. 94.

11. Stuart Nagel, "Sociometric Relations among American Courts," *Southwestern Social Science Quarterly* 43 (September 1962): 136–42; also in Nagel, *The Legal*

Process from a Behavioral Perspective (Homewood, Ill.: Dorsey Press, 1969), pp. 59–68.

12. Alan Schechter, "Impact of Open Housing Laws on Suburban Realtors," *Urban Affairs Quarterly* 8 (June 1973): 439–65.

13. Neil T. Romans, "The Role of State Supreme Courts in Judicial Policy-Making: *Escobedo, Miranda* and the Use of Judicial Impact Analysis," *Western Political Quarterly* 27 (March 1974): 38–59.

14. Bradley C. Canon, "Organizational Contumacy in the Transmission of Judicial Policies: The *Mapp, Escobedo, Miranda,* and *Gault* Cases," *Villanova Law Review* 20 (November 1974): 69.

15. Charles Hamilton, *The Bench and the Ballot: Southern Federal Judges and Black Voters* (New York: Oxford University Press, 1973).

16. Kenneth N. Vines, "Southern State Supreme Courts and Race Relations," *Western Political Quarterly* 18 (March 1965): 5–18.

17. J. W. Peltason, *Fifty-Eight Lonely Men: Southern Federal Judges and School Desegregation* (Urbana: University of Illinois Press, 1971 [1961]), pp. 245–46. See also Kenneth Vines, "Federal District Judges and Race Relations Cases in the South," *Journal of Politics* 26 (May 1964): 337–57; Michael W. Giles and Thomas G. Walker, "Judicial Policy-Making and Southern School Segregation," *Journal of Politics* 37 (November 1975): 917–36.

18. Wayne LaFave and Frank Remington, "Controlling the Police: The Judge's Role in Making and Reviewing Law Enforcement Decisions," *Michigan Law Review* 63 (April 1965): 1005.

19. Baum, "Implementation of Judicial Decisions," p. 95.

20. Thomas E. Barth, "Perception and Acceptance of Supreme Court Decisions at the State and Local Level," *Journal of Public Law* 17 (1968): 308–50.

21. See Gerald Caplan, "The Police Legal Adviser," *Journal of Criminal Law, Criminology, and Police Science* 58 (September 1967): 303-309, and Frank Carrington, "Speaking for the Police," ibid., 61 (June 1970): 244–79.

22. See David L. Grey, *The Supreme Court and the Mass Media* (Evanston, Ill.: Northwestern University Press, 1968), on this point and others discussed here.

23. Chester Newland, "Press Coverage of the United States Supreme Court," *Western Political Quarterly* 17 (March 1964): 15–36.

24. See David W. Leslie, "The Supreme Court in the Media: A Content Analysis," paper presented to the International Communication Association, 1976, and David W. Leslie and D. Brock Hornby, *The Supreme Court in the Media: A Theoretical and Empirical Analysis* (Final Technical Report to National Science Foundation, 1976, Grant GS 38113).

25. Neal Milner, *The Court and Local Law Enforcement: The Impact of Miranda* (Beverly Hills, Calif.: Sage Publications, 1971), p. 47.

26. Ibid., pp. 52, 226.

27. Larry Berkson, "The United States Supreme Court and Small-Town Police Officers: A Study in Communication," unpublished manuscript, 1970.

28. Data from Illinois and Massachusetts in this and the following paragraph from Stephen L. Wasby, *Small Town Police and the Supreme Court;* large-city data from a study by Bradley C. Canon, also reported in ibid; see p. 96, and chap. 5 and 6, pp. 119–98, passim.

29. Ibid., pp. 104–108, 96–100.
30. Stuart Nagel, "Testing the Effects of Excluding Illegally Seized Evidence," *Wisconsin Law Review* 1965 (Spring): 283–310; also in Nagel, *The Legal Process from a Behavioral Perspective*, pp. 294–320.
31. Milner, *The Court and Local Law Enforcement*, p. 52.
32. Michael J. Murphy, "The Problem of Compliance by Police Departments," *Texas Law Review* 44 (1966): 941.
33. Wayne R. LaFave, "Improving Police Performance through the Exclusionary Rule: Part II: Defining the Norms and Training the Police," *Missouri Law Review* 30 (Fall 1965): 594–95.
34. David Manwaring, "The Impact of *Mapp v. Ohio*," *The Supreme Court as Policy-Maker: Three Studies on the Impact of Judicial Decisions*, edited by David Everson (Carbondale, Ill.: Public Affairs Research Bureau, Southern Illinois University, 1968), p. 26.
35. Bradley C. Canon, "Reactions of State Supreme Courts to a U.S. Supreme Court Civil Liberties Decision," *Law & Society Review* 8 (Fall 1973): 126–217.
36. See Canon, "Organizational Contumacy," pp. 50–79.
37. *Bivens v. Six Unknown Federal Narcotics Agents*, 403 U. S. 388 at 416 (1971).
38. Dallin H. Oaks, "Studying the Exclusionary Rule in Search and Seizure," *University of Chicago Law Review* 37 (Summer 1970): 667, 655.
39. Bradley C. Canon, "Is the Exclusionary Rule in Failing Health? Some New Data and a Plea Against a Precipitous Conclusion," *Kentucky Law Journal* 62 (1973–1974): 708–709.
40. Michael Wald et al., "Interrogations in New Haven: The Impact of Miranda," *Yale Law Journal* 76 (July 1967): 1519–1648; Richard J. Medalie et al., "Custodial Police Interrogation in Our Nation's Capital: The Attempt to Implement Miranda," *Michigan Law Review* 66 (May 1968): 1347–1422; Albert J. Reiss, Jr., and Donald J. Black, "Interrogation and the Criminal Process," *The Annals*, vol. 374 (November 1967), pp. 47–57.
41. John Griffiths and Richard E. Ayres, "A Postscript to the *Miranda* Project; Interrogation of Draft Protestors," *Yale Law Journal* 76 (December 1967): 300–319.
42. Milner, *The Court and Local Law Enforcement*, p. 229.
43. Ibid., p. 217.
44. Barton L. Ingraham, "The Impact of Argersinger—One Year Later," *Law & Society Review* 8 (Summer 1975): 616.
45. Sheldon Krantz et al., "The Right to Counsel in Criminal Cases: The Mandate of *Argersinger v. Hamlin*," Summary Report (Washington, D.C.: National Institute of Law Enforcement and Criminal Justice, Law Enforcement Assistance Administration, Department of Justice, 1976), pp. 2–3.
46. Gordon Patric, "The Impact of a Court Decision: Aftermath of the McCollum Case," *Journal of Public Law* 6 (Fall 1967): 455–65.
47. Frank J. Sorauf, "Zorach v. Clauson: The Impact of a Supreme Court Decision," *American Political Science Review* 53 (September 1959): 777–91.
48. See Richard Johnson, *The Dynamics of Compliance* (Evanston, Ill.: Northwestern University Press, 1967). Among the other important studies are those by Donald

Reich, "The Impact of Judicial Decision-Making: The School Prayer Cases," *The Supreme Court as Policy-Maker,* edited by D. Everson, pp. 44–81, and Robert Birkby, "The Supreme Court and the Bible Belt: Tennessee Reaction to the Schempp Decision," *Midwest Journal of Political Science* 10 (August 1966): 304–19.

49. H. Frank Way, Jr., "Survey Research on Judicial Decisions: The Prayer and Bible Reading Cases," *Western Political Quarterly* 21 (June 1968): 191.

50. Kenneth Dolbeare and Phillip Hammond, *The School Prayer Decisions: From Court Policy to Local Practice* (Chicago: University of Chicago Press, 1971).

51. Muir, *Prayer in the Public Schools.*

52. *Harris v. New York,* 401 U.S. 222 (1971), *Oregon v. Hass,* 420 U.S. 714 (1975), and *Michigan v. Mosley,* 96 S.Ct. 321 (1975).

53. Baum, "Implementation of Judicial Decisions," p. 93.

54. Milner, *The Court and Local Law Enforcement,* pp. 230–31.

55. Edwin Lemert, *Social Action and Legal Change: Revolution within the Juvenile Court* (Chicago: Aldine, 1970), pp. 221–22.

56. Earl Black, "Southern Governors and Political Change: Campaign Stances on Racial Segregation and Economic Development," *Journal of Politics* 33 (August 1971): 709.

57. This paragraph draws on Charles S. Bullock III and Harrell R. Rodgers, Jr., "Communications between the Regulated and the Regulators: A Case of Distortion," paper presented to the International Communication Association, 1976. See also Harrell R. Rodgers, Jr., and Charles S. Bullock III, *Coercion to Compliance* (Lexington, Mass.: Lexington Books, 1976).

58. Michael Ban, "The Impact of *Mapp V. Ohio* on Police Behavior," paper presented to the Midwest Political Science Association, 1973, p. 33 (based on Boston and Cincinnati).

59. For a discussion of the police work situation, see Neal A. Milner, "Supreme Court Effectiveness and Police Organization," *Law and Contemporary Problems* 36 (Autumn 1971): 467–87.

60. Leslie and Hornby, *The Supreme Court in the Media,* pp. 142–43.

61. James P. Levine, "Constitutional Law and Obscene Literature: An Investigation of Bookseller Censorship Practicies," *The Impact of Supreme Court Decisions: Empirical Studies,* edited by Theodore L. Becker (New York: Oxford University Press, 1969), pp. 129–48.

62. Harrell R. Rodgers, Jr., and Charles S. Bullock III, "Law Abidingness as a Guide to Official Decision-Making: Techniques of Neutralization," unpublished manuscript; see also Rodgers and Edward B. Lewis, "Political Support and Compliance Attitudes: A Study of Adolescents," *American Political Quarterly* 2 (January 1974): 61–77; Rodgers and George Taylor, "Pre-Adult Attitudes Toward Legal Compliance," *Social Science Quarterly* 51 (December 1970): 539–51; and Don W. Brown, "Cognitive Development and Willingness to Comply with Law," *American Journal of Political Science* 18 (August 1974): 583–94.

BIBLIOGRAPHY

The reader who wishes to explore further some of the points raised in this book might profitably start with some of the books listed below. Included are those which formed the basis of this book along with a number of others of both general and specific coverage. Articles in journals are not included; for those, the reader is referred to the notes at the end of each chapter.

Abraham, Henry J. *The Judicial Process,* 2nd ed. New York: Oxford University Press, 1968.
———. *Justices & Presidents: A Political History of Appointments to the Supreme Court.* New York: Oxford University Press, 1974.
Becker, Theodore L., ed. *The Impact of Supreme Court Decisions: Empirical Studies.* New York: Oxford University Press, 1969. 2nd ed., 1973, edited by Theodore Becker and Malcolm Feeley.
Bickel, Alexander. *The Least Dangerous Branch.* Indianapolis: Bobbs-Merrill, 1962.
Black, Charles, L., Jr. *The People and the Court: Judicial Review in a Democracy.* Englewood Cliffs, N.J.: Prentice-Hall, 1960.
Bullock, Charles III, and Harrell R. Rodgers, Jr. *Law and Social Change.* New York: McGraw-Hill, 1972.
Casper, Jonathan. *Lawyers Before the Warren Court: Civil Liberties and Civil Rights,* 1957–66. Urbana: University of Illinois Press, 1972.
Chase, Harold W. *Federal Judges: The Appointing Process.* Minneapolis: University of Minnesota Press, 1972.
Commission on Revision of the Federal Court Appellate System. *Structure and Internal Procedures: Recommendations* Washington, D.C., 1975.
D'Amato, Anthony, and Robert O'Neil. *The Judiciary and Vietnam.* New York: St. Martin's Press, 1972.

Danelski, David. *A Supreme Court Justice Is Appointed.* New York: Random House, 1964.

Dean, Howard. *Judicial Review and Democracy.* New York: Random House, 1966.

Dolbeare, Kenneth, and Phillip Hammond. *The School Prayer Decisions: From Court Policy to Local Practice.* Chicago: University of Chicago Press, 1971.

Eisenstein, James. *Politics and the Legal Process.* New York: Harper and Row, 1973.

Fish, Peter Graham. *The Politics of Federal Judicial Administration.* Princeton, N.J.: Princeton University Press, 1973.

Friedman, Leon, ed. *Argument: The Oral Argument Before the Supreme Court in Brown v. Board of Education of Topeka,* 1952–1955 (New York: Chelsea House, 1969).

Friedman, Leon, ed. *Obscenity: The Complete Oral Arguments Before the Supreme Court in the Major Obscenity Cases* (New York: Chelsea House, 1970).

Goldman, Sheldon, and Thomas Jahnige. *The Federal Courts as a Political System,* 2nd ed. New York: Harper and Row, 1976.

Graham, Fred. *The Self-Inflicted Wound.* New York: Macmillan, 1970.

Grey, David L. *The Supreme Court and the Mass Media.* Evanston, Ill.: Northwestern University Press, 1968.

Grossman, Joel. *Lawyers and Judges: The ABA and the Politics of Judicial Selection.* New York: John Wiley, 1965.

———, and Joseph Tanenhaus, eds., *Frontiers of Judicial Research.* New York: John Wiley, 1969.

———, and Richard Wells, *Constitutional Law and Judicial Policy-Making.* New York: John Wiley, 1972.

Hamilton, Charles. *The Bench and the Ballot: Southern Federal Judges and Black Voters.* New York: Oxford University Press, 1973.

Harvard Law Review. Annual November issue on Supreme Court.

Jacob, Herbert. *Justice in America: Courts, Lawyers, and the Judicial Process,* 2nd ed. Boston: Little, Brown, 1972.

Jahnige, Thomas P., and Sheldon Goldman, eds. *The Federal Judicial System: Readings in Process and Behavior.* New York: Holt, Rinehart, and Winston, 1968.

Johnson, Richard. *The Dynamics of Compliance.* Evanston, Ill.: Northwestern University Press, 1967.

Kelly, Alfred H., and Winfred A. Harbison, *The American Constitution: Its Origins and Development,* 4th ed. New York: W. W. Norton, 1970.

Krislov, Samuel. *The Supreme Court in the Political Process.* New York: Macmillan, 1965.

———et al., eds., *Compliance and the Law.* Beverly Hills, Calif.: Sage Publications, 1972.

Kurland, Philip, ed. *Supreme Court Review.* Single volume annually. Chicago: University of Chicago Press.

Miller, Arthur S. *The Supreme Court and American Capitalism.* New York: Free Press, 1970.

Milner, Neal. *The Supreme Court and Local Law Enforcement: The Impact of Miranda.* Beverly Hills, Calif.: Sage Publications, 1971.

Morgan, Donald G. *Congress and the Constitution: A Study of Responsibility.* Cambridge, Mass.: Harvard University Press, 1966.

Muir, William K. *Prayer in the Public Schools: Law and Attitude Change.* Chicago: University of Chicago Press, 1967.

Murphy, Walter F. *Congress and the Court.* Chicago: University of Chicago Press, 1962.

———. *Elements of Judicial Strategy.* Chicago: University of Chicago Press, 1964.

———, and Joseph Tanenhaus. *The Study of Public Law.* New York: Random House, 1972.

———, and C. Herman Pritchett, eds. *Courts, Judges, and Politics,* 2nd ed. New York: Random House, 1974.

Nagel, Stuart. *The Legal Process from a Behavioral Perspective.* Homewood, Ill.: Dorsey Press, 1969.

Peltason, Jack W. *Fifty-Eight Lonely Men: Southern Federal Judges and School Desegregation.* Urbana: University of Illinois Press, 1961/1971.

Pritchett, C. Herman. *The American Constitutional System,* 4th ed. New York: McGraw-Hill, 1976.

———. *The Roosevelt Court: A Study in Judicial Politics and Values,* 1937–1947. New York: Macmillan, 1948.

———. *The Vinson Court and Civil Liberties.* Chicago: University of Chicago Press, 1963.

———, and Alan F. Westin, eds. *The Third Branch of Government.* New York: Harcourt, Brace, and World, 1963.

Richardson, Richard J., and Kenneth N. Vines. *The Politics of Federal Courts.* Boston: Little, Brown, 1970.

Rodgers, Harrell R., Jr., and Charles S. Bullock III. *Coercion to Compliance.* Lexington, Mass.: Lexington, 1967.

Rohde, David, and Harold Spaeth. *Supreme Court Decision-Making.* San Francisco: W. H. Freeman, 1975.

Rossiter, Clinton. *The Supreme Court and the Commander in Chief.* Ithaca, N. Y.: Cornell University Press, 1951.

Schmidhauser, John R. *The Supreme Court: Its Politics, Personalities and Procedures.* New York: Holt, Rinehart, and Winston, 1961.

———, ed. *Constitutional Law in the Political Process.* Chicago: Rand McNally, 1963.

Schubert, Glendon. *Constitutional Politics.* New York: Holt, Rinehart, and Winston, 1960.

———. *The Constitutional Polity.* Boston: Boston University Press, 1970.

———. *The Judicial Mind.* Evanston, Ill.: Northwestern University Press, 1965.

———. *The Judicial Mind Revisited: Psychometric Analysis of Supreme Court Ideology.* New York: Oxford University Press, 1974.

———. *Judicial Policy-Making,* 2nd ed. Glenview, Ill.: Scott, Foresman, 1974.

———. *The Presidency in the Courts.* Minneapolis: University of Minnesota Press, 1957.

———. *Quantitative Analysis of Judicial Behavior.* Glencoe, Ill.: Free Press, 1959.

———, ed. *Judicial Behavior: A Reader in Theory and Research.* Chicago: Rand McNally, 1964.

———, ed. *Judicial Decision-Making.* New York: Free Press, 1963.

Scigliano, Robert. *The Supreme Court and the Presidency.* New York: Free Press, 1971.

———, ed. *The Courts: A Reader in the Judicial Process.* Boston: Little, Brown, 1962.

Shapiro, Martin. *Law and Politics in the Supreme Court.* New York: Free Press, 1964.

———. *The Supreme Court and the Administrative Agencies.* New York: Free Press, 1968.

Sheldon, Charles H., *The American Judicial Process: Models and Approaches.* New York: Dodd, Mead, 1974.

Shogan, Robert. *A Question of Judgment: The Fortas Case and the Struggle for the Supreme Court.* Indianapolis: Bobbs-Merrill, 1972.

Simon, James F. *In His Own Image: The Supreme Court in Richard Nixon's America.* New York: David McKay, 1973.

Sorauf, Frank, J. *The Wall of Separation: The Constitutional Politics of Church and State.* Princeton, N.J.: Princeton University, 1976.

Sprague, John D. *Voting Patterns of the United States Supreme Court.* Indianapolis: Bobbs-Merrill, 1968.

Vose, Clement. *Caucasians Only: The Supreme Court, the NAACP, and the Restrictive Covenant Cases.* Berkeley: University of California Press, 1959.

Wasby, Stephen L. *Continuity and Change: From the Warren Court to the Burger Court.* Pacific Palisades, Calif.: Goodyear Publishing, 1976.

———. *The Impact of the United States Supreme Court: Some Perspectives.* Homewood, Ill.: Dorsey Press, 1970.

———. *Small Town Police and the Supreme Court: Hearing the Word.* Lexington, Mass.: Lexington Books, 1976.

———, Anthony D'Amato, and Rosemary Metrailer. *Desegregation from Brown to Alexander: An Exploration of Supreme Court Strategies.* Carbondale, Ill.: Southern Illinois University Press, 1977.

Westin, Alan. *The Anatomy of a Constitutional Law Case.* New York: Macmillan, 1958.

———, ed. *An Autobiography of the Supreme Court: Off-the-Bench Commentary by the Justices.* New York: Macmillan, 1963.

Wheeler, Russell, and Howard R. Whitcomb, eds., *Perspectives on Judicial Administration: Readings in Court Management and the Administration of Justice.* Englewood Cliffs, N.J.: Prentice-Hall, 1976.

TABLE OF CASES*

Abington School District v. Schempp, 394 U.S. 203 (1963), 19, 28*n*10, 231–32, 233
Ableman v. Booth, 21 How. 506 (1859), 40
Abrams v. United States, 394 U.S. 165 (1969), 27*n*2
Afroyim v. Rusk, 387 U.S. 254 (1967), 211*n*1
Albemarle Paper Co. v. Moody, 422 U.S. 405 (1975), 29*n*27
Albertson v. Subversive Activities Control Board, 382 U.S. 70 (1965), 212*n*3
Alderman v. United States, 394 U.S. 165 (1969), 203
Aldinger v. Howard, 96 S.Ct. 2413 (1976), 134*n*29
Alexander v. Holmes County, 396 U.S. 19 (1969), 153
Allee v. Medrano, 416 U.S. 802 (1974), 125
Allegheny-Ludlum Steel Corp., United States v.
Almeida-Sanchez v. United States, 413 U.S. 266 (1973), 29*n*35
Alyeska Pipeline Co. v. Wilderness Society, 421 U.S. 240 (1975), 117
American Federation of Government Employees v. Phillips, 358 F.Supp. 60 (D.D.C. 1973), 213*n*23
Andresen v. Maryland, 96 S.Ct. 2737 (1976), 29*n*20, 29*n*35
Aptheker v. Secretary of State, 378 U.S. 500 (1964), 212*n*3
Argersinger v. Hamlin, 407 U.S. 25 (1972), 30*n*38, 229–30
Arizona v. New Mexico, 96 S.Ct. 1845 (1976), 81*n*5
Ash, United States v.
Ashwander v. Tennessee Valley Authority, 297 U.S. 288 (1936), 31*n*63
Askew v. American Waterways Operators, 411 U.S. 325 (1973), 53*n*8
Atkins v. United States, 96 S.Ct. 3162 (1976), 85
Automobile Workers v. Wisconsin Board, 336 U.S. 245 (1949), 28*n*20

Bailey v. Drexel Furniture Co., 259 U.S. 20 (1922), 27*n*3
Bailey v. Weinberger, 419 U.S. 953 (1974), 155*n*8
Baker v. Carr, 369 U.S. 186 (1962), 26, 31*n*62, 119, 221–22
Barlow v. Collins, 397 U.S. 159 (1970), 133*n*7
Barrows v. Jackson, 346 U.S. 249 (1953), 133*n*5
Beckwith v. United States, 96 S.Ct. 1612 (1976), 30*n*37
Bellotti v. Baird, 96 S.Ct. 2857 (1976), 134*n*25
Berger v. New York, 388 U.S. 41 (1967), 28*n*12, 193
Bishop v. Wood, 96 S.Ct. 2074 (1976), 29*n*23
Bivens v. Six Unknown Federal Narcotics Agents, 403 U.S. 388 (1971), 240*n*37
Blount v. Rizzi, 400 U.S. 419 (1971), 212*n*6
Bond v. Floyd, 385 U.S. 116 (1966), 119
Boyd v. United States, 116 U.S. 616 (1886), 10
Brady v. United States, 397 U.S. 742 (1970), 30*n*39, 134*n*32
Branzburg v. Hayes, 408 U.S. 665 (1972), 29*n*33, 189
Brewster, United States v.
Briggs v. Elliott: see Brown v. Board of Education

*When the case name is not mentioned in the text, the reference is to the Notes at the end of each chapter, from which the reader can refer to the appropriate place in the text.

INDEX